T0213296

SpringerBriefs in Molecular Science

Chemistry of Foods

Series Editor

Salvatore Parisi, Al-Balqa Applied University, Al-Salt, Jordan

The series Springer Briefs in Molecular Science: Chemistry of Foods presents compact topical volumes in the area of food chemistry. The series has a clear focus on the chemistry and chemical aspects of foods, topics such as the physics or biology of foods are not part of its scope. The Briefs volumes in the series aim at presenting chemical background information or an introduction and clear-cut overview on the chemistry related to specific topics in this area. Typical topics thus include:

- Compound classes in foods—their chemistry and properties with respect to the foods (e.g. sugars, proteins, fats, minerals, …)
- Contaminants and additives in foods—their chemistry and chemical transformations
- Chemical analysis and monitoring of foods
- Chemical transformations in foods, evolution and alterations of chemicals in foods, interactions between food and its packaging materials, chemical aspects of the food production processes
- Chemistry and the food industry—from safety protocols to modern food production

The treated subjects will particularly appeal to professionals and researchers concerned with food chemistry. Many volume topics address professionals and current problems in the food industry, but will also be interesting for readers generally concerned with the chemistry of foods. With the unique format and character of SpringerBriefs (50 to 125 pages), the volumes are compact and easily digestible. Briefs allow authors to present their ideas and readers to absorb them with minimal time investment. Briefs will be published as part of Springer's eBook collection, with millions of users worldwide. In addition, Briefs will be available for individual print and electronic purchase. Briefs are characterized by fast, global electronic dissemination, standard publishing contracts, easy-to-use manuscript preparation and formatting guidelines, and expedited production schedules.

Both solicited and unsolicited manuscripts focusing on food chemistry are considered for publication in this series. Submitted manuscripts will be reviewed and decided by the series editor, Prof. Dr. Salvatore Parisi.

To submit a proposal or request further information, please contact Tanja Weyandt, Publishing Editor, via tanja.weyandt@springer.com or Prof. Dr. Salvatore Parisi, Book Series Editor, via drparisi@inwind.it or drsalparisi5@gmail.com

More information about this subseries at http://www.springer.com/series/11853

Maria Anna Coniglio · Cristian Fioriglio ·
Pasqualina Laganà

Non-Intentionally Added Substances in PET-Bottled Mineral Water

 Springer

Maria Anna Coniglio
Department of Medical, Surgical
Sciences and Advanced Technologies
"G.F. Ingrassia"
University of Catania
Catania, Italy

Cristian Fioriglio
Police Headquarters
General Prevention & Public Aid Office
Catania, Italy

Pasqualina Laganà
Department of Biomedical and Dental
Sciences and Morphofunctional Imaging
University of Messina
Messina, Italy

ISSN 2191-5407   ISSN 2191-5415   (electronic)
SpringerBriefs in Molecular Science
ISSN 2199-689X   ISSN 2199-7209   (electronic)
Chemistry of Foods
ISBN 978-3-030-39133-1   ISBN 978-3-030-39134-8   (eBook)
https://doi.org/10.1007/978-3-030-39134-8

This Springer imprint is published by the registered company Springer Nature Switzerland AG
The registered company address is: Gewerbestrasse 11, 6330 Cham, Switzerland

Preface

This book deals with polyethylene terephthalate (PET) bottled water, whose global market size is expected to reach around USD 350 billion by 2021, following 10% year-on-year growth. Our goal is to define and formulate a comprehensive analysis of risk associated with drinking PET-bottled water.

In recent years, concerns are rising about the safety of PET foods packaging due to the possible migration of chemical compounds from PET to the water contained into the bottle, which may pose health risk to consumers. These chemicals are called 'NIAS', acronym of non-intentionally added substances, and they are supposed to have potential estrogenic and/or anti-androgenic activities and to be cancerogenic or toxic to humans. Our approach is to report data from the international scientific literature examining what happens when the polymer undergoes specific conditions, such as exposure to ultraviolet light/high temperature, ageing or humidity.

The introductory chapter highlights the relationship between water and health. Taking into consideration the recommended daily water intake, the importance of good hydration is argued. The second chapter deals with the global bottled market. The demand for bottled water is witnessing significant growth due to organoleptic and health-related reasons. Concerns about tap water risks may contribute to bottled water consumption. In particular, demographic and socio-economic variables as well as information provided by the mass media are argued as factors related to the consumers' attitudes towards bottled drinking waters. Furthermore, European Union (EU) and United States of America (U.S.) legislation on bottled waters is examined. The third chapter deals with the fundamentals of PET chemistry and the elementary PET-bottled manufacturing steps. It is a guide to the reader for the comprehension of what NIAS are and how they can migrate from the wall of PET bottle to the water inside it. For this reason, special emphasis is given to the additives used during the synthesis of the polymer. In addition, specific EU and U.S. regulations on plastic materials and articles intended to come into contact with foodstuffs are examined. Finally, the last chapter describes NIAS as a result of the interactions between different ingredients in the packaging materials, as well as the degradation processes and the impurities present in the raw materials used for their production. Phthalates, aldehydes and volatile organic compounds (VOC) are examined. Although some of them may be cancerogenic or toxic to humans, it is likely that in the majority of cases, due to their very low levels, these substances will not be of any health concern.

We are conscious that the book is not exhaustive of all the current topics related to PET and NIAS. Nonetheless, because with the approach described we attempt to answer questions not only of 'how' NIAS migrate but also 'why' they migrate from PET into the bottled water, we hope that this book can be a useful compendium of data that are currently in literature.

We acknowledge with pleasure the colleagues who helped us in our efforts. Foremost, we thank Doctor Stefano Melada who reviewed the entire manuscript and provided invaluable help and advice on the content of the book. We further acknowledge the constructive discussions and suggestions offered. We would also thank Carmelo Parisi, currently a student at the Liceo Scientifico Stanislao Cannizzaro, Palermo, Italy, for some of realised figures in this book. We also thank Prof. Salvatore Parisi (Al-Balqa Applied University, Jordan) for his thorough editing of the manuscript, which contributed greatly to the final quality of the book.

Catania, Italy   Maria Anna Coniglio
Catania, Italy   Cristian Fioriglio
Messina, Italy   Pasqualina Laganà

Contents

Chapter 1
Water and Health

Abstract Water is essential for all functions of the body: metabolism, substrate transport across membranes, cellular homeostasis, temperature regulation and circulatory function. A water intake which balances losses and which assures adequate hydration of body tissues is essential for health. Water is consumed from different sources, which include water produced by oxidative processes in the body, drinking water, beverages and foods. Drinking water and beverages represent 70–80% of total fluid intake, while water coming from food represents about 20–30% of the total intake. Today, neither upper nor lower water consumption limits have been clearly linked to a specific benefit or risk, and most of the guidelines for total water intake are based on median population intake.

Keywords Dehydration · Drinking water · Extracellular water · Fat-free mass · Hypohydration · Intracellular water · Water intake

Abbreviations

AI	Adequate Intake
CO_2	Carbon dioxide
CDC	Centers for Disease Control and Prevention
EFSA	European Food Safety Authority
EU	European Union
IOM	Institute of Medicine
TWI	Total Water Intake
U.S.	United States of America

1.1   Water Distribution in the Body

Water is the main constituent of the human body. The quantity of water in the human body varies among individuals and is primarily determined by age, gender and body composition. While the water content accounts for about 74% of total body mass in infants, it decreases at 55–60% in the adult. In the elderly, it is further reduced to 51% in men and 45% in women (Nicolaidis 1998). Total body water is also determined by gender, accounting about 60% of body weight in adult males and 50–55% in females (EFSA 2010; IOM 2004). Finally, taking into consideration the body composition, the body fat relative mass directly influences total body water. In fact, the water content in lean body mass is constant, and it accounts for about 73% (Peronnet et al. 2012; Sawka et al. 2005; Wang et al. 1999). This fact explains the influence of age and gender on total body water because old people and women have lower fat-free mass than men and infants (IOM 2004; Marieb and Hoehn 2007).

Total body water is subdivided into intracellular water and extracellular water. It is estimated that almost two-thirds of total body fluid is intracellular, while the remaining third is extracellular fluid which, in turn, is subdivided into interstitial liquids (14% of body weight), plasma (4% of body weight), transcellular water (1% of body weight) and lymphatic liquids (1% of body weight). Intracellular water is the index of metabolically active mass or 'fat-free mass', which is greater in the male sex: men: 60%, women: 50% (Bedogni et al. 1996). Transcellular fluid is contained in body cavities, such as peritoneal and pleural fluids, cerebral spinal and synovial fluid (Nissensohn et al. 2015). Finally, body fluids containing the largest amount of water are cerebrospinal fluid, bone marrow fluid (99%), blood plasma (85%) and brain (75%) (Fellin 2003).

The water content of various organs depends on their composition, ranging from 83% in blood to only 10% in human adipose tissue (Fig. 1.1).

1.2   The Importance of Good Hydration

Water is involved in many body functions (Figs. 1.1 and 1.2). In particular, it helps to regulate the internal body temperature; it supports the digestion of food, as a solvent it transports and distributes oxygen, nutrients, metabolites and hormones into cells through the circulatory system; and it helps to eliminate waste and toxins from the body (Kleiner 1999).

Water homeostasis is essential for hydroelectrolytic balance, acid–base balance and thermal balance. Dehydration means essentially water loss in human beings, while hypohydration is for a diminution of the aqueous content in the human organism as result of dehydration phenomena (EFSA 2010). It is well known that water reduction of 2% of body weight alters thermoregulation and plasma volume; 7% reduction may cause hallucinations; dehydration of 10% may cause death (Del Toma 1995).

Water composition of tissues and organs by weight (%)

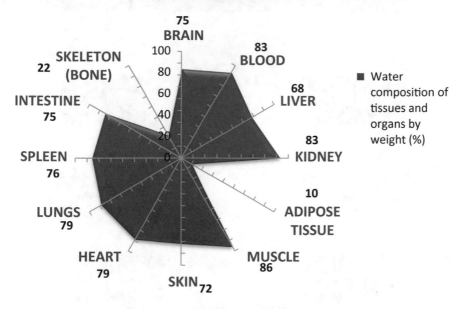

Fig. 1.1 Water composition of tissues and organs by weight. Adapted data (Pivarnik and Palmer 1994)

Drinking water is important to maintaining the function of every system in the body. Thereby, there is increasing evidence that mild dehydration plays a role in the development of various diseases.

The performance of both physical and mental tasks can be adversely affected by dehydration. There is growing evidence that the effects of dehydration on exercise performance may be mediated by effects on the central nervous system. This seems to involve serotonergic and dopaminergic functions. Recent evidence suggests that the integrity of the blood–brain barrier may be compromised by combined heat stress and dehydration, and this may play a role in limiting performance in the heat.

Maintaining good hydration status may prevent urolithiasis and may be beneficial in treating urinary tract infections (Manz and Wentz 2005). Each patient with nephrolithiasis should increase his fluid intake to achieve a daily urine output of 2 litres (Ticinesi et al. 2017).

Good hydration is also associated with lower rates of death from coronary heart disease in middle-aged and elderly people, hypertension, venous thromboembolism and cerebral infarct. Water intake reduces heart rate and increases blood pressure in both normotensive and hypertensive individuals (Callegaro et al. 2007). In fact, blood volume, blood pressure and heart rate are closely linked and influenced by water balance (Shirreffs and Maughan 1998).

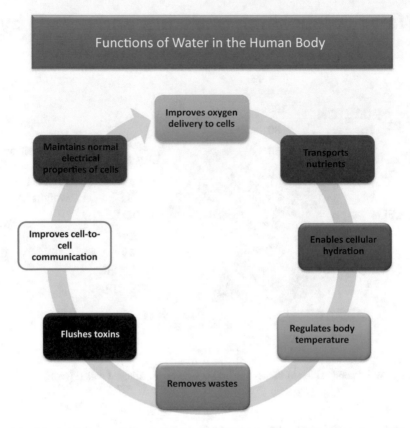

Fig. 1.2 Functions of water in human body. This picture has been realised by Carmelo Parisi, currently a student at the Liceo Scientifico Stanislao Cannizzaro, Palermo, Italy

Adequate fluid intake has also been associated with benefits in bronchopulmonary disorders. The movement of fluid between the airspaces, interstitium and vascular compartments in the lungs seems to play an important role in the maintenance of hydration and protection of the lung epithelium and it seems to significantly contribute to the airway clearance. In particular, there would be a relationship between hydration status and bronchopulmonary disorders like bronchitis and asthma. Asthma is characterised by chronic airway inflammation and episodic airway obstruction. Data indicate an association between exercise-induced asthma and humidity of inspired air. For this reason, some expert groups recommend sufficient hydration as a complementary asthma therapy. On the other side, there are no reliable confirmations concerning bronchopulmonary disorders as determined totally or partially by mild dehydration (Kalhoff 2003).

Good hydration is considered fundamental also for the management of type 1 diabetes because it limits the development of diabetic ketoacidosis during insulin deficiency (Manz and Wentz 2005). Meta-analyses of prospective cohort studies,

studies with animal models and interventional studies suggest a probable positive effect of drinking water and mineral water on glycemic parameters (Naumann et al. 2017).

Finally, adequate hydration can prevent other different diseases such as hypertension, venous thromboembolism and other problems. Good hydration may also be associated with the improvement of the skin. The skin contains approximately 30% water, and it is important in maintaining body water levels. Serious levels of dehydration can be reflected in reduced skin turgor, while an adequate water intake can improve skin thickness and density, as well as skin hydration (Mac-Mary et al. 2006).

The study of the effects of hydration status on cognitive function represents a relatively new area of research. Several dose–response studies on human cognitive function report that dehydration levels of 1% may adversely affect cognitive performance, and other problems concerning psychomotor, short-term memory and cognitive performance at least can be observed in similar situations. Anyway, additional research is needed because not all studies find behavioural effects in this range (Cian et al. 2000; Grandjean 2007; Lieberman 2007; Neave et al. 2001; Szinnai et al. 2005).

On the other hand, there is a rich scientific literature regarding hydration status and physical function. Dehydration during physical activity in the heat provokes greater performance decrements than similar activity in cooler conditions. Dehydration levels can be critical in certain situations (Murray 2007).

1.3 Beverages, Food and Water Intake

Water is consumed as drinking water, beverages and food moisture. Drinking water and beverages provide 800–1,500 ml of water daily intake (on average 70–80% of the total fluid intake), while foods provide 500–900 ml of water daily intake (20–30%). The amount of water in foods depends on the cooking method, as well as on their composition: water content is 10–15% in dry foods and about 90% in fruits and vegetables (EFSA 2010). Soups are the category that contains the second highest level of water after fruits and vegetables, with values ranging between 82 and 95% in water, depending on recipes. On the contrary, some dry products, such as biscuits and chocolate, have aqueous amounts below 5% (McCance RA 2002).

Another source of water for the body includes water that results from oxidation of organic materials. This 'metabolic' water accounts for about 350 ml/day, and it can be recognised as an important component of the total water income of animals. Metabolic water results from macronutrient oxidation into water and carbon dioxide (CO_2). Nutrients differ in their yield of metabolic water, with fats providing the most part of water amount per gram. It has been calculated (Fig. 1.3) that oxidation of 100 g lipids produces 107 ml of water; oxidation of 100 g carbohydrates produces 60 ml; and the oxidation of 100 g proteins produces 40 ml (Del Toma 1995; IOM 2004).

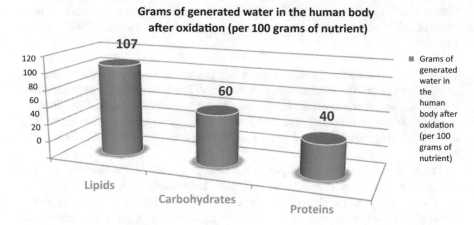

Fig. 1.3 The metabolic water production per gram in the human body depends on nutrients. On these bases, it may be inferred that nutrients can be easily discriminated depending on the amount of water which can be produced by means of simple biological oxidation (Del Toma 1995; IOM 2004)

1.4 Recommended Daily Water Intake

We lose water every day through urine (1,500 ml/day), skin (*perspiratio insensibilis*, 900 ml/day), faeces (100 ml/day) and lungs (moist gases). But we also increase our extraction of water by drinking alcohol, beer, coffee, tea, caffeinated cola drinks and taking some drugs. In fact, all are diuretics. Since endogenous water is not enough to counterbalance water loss (about 2,150–2,900 ml/day), an adequate exogenous water intake is necessary. Thus, water must be restored through diet.

Water intake from beverages and foods is defined as total water intake, while the sum of total water intake and oxidation water constitutes the so-called total available water. The Centers for Disease Control and Prevention (CDC 2016) states that '*There is no recommendation for how much plain water adults and youth should drink daily*'.[1] However, there is general consensus on the evidence that water requirement varies according to age, gender, environmental conditions (e.g. temperature and humidity), physical activity, water content in foods and health conditions. Exercise, for example, increases water waste. It has been observed that during physical activity only 50% of water lost through sweat is replaced in response to the sensation of thirst. Moreover, water requirement increases in all pathologic conditions resulting in major water loss, such as diarrhoea, vomit, hyperpyrexia or abundant perspiration.

In order to allow metabolic processes to function correctly, it is important to maintain a correct 'fluid balance': this term is used to describe the balance of the input and output of fluids in the body (Welch 2010). Most of the components of fluid balance are controlled by homeostatic mechanisms, which are activated with deficits

---

[1]For this and other information, the following web site can be accessed: https://www.cdc.gov/nutrition/data-statistics/plain-water-the-healthier-choice.html.

or excesses of water. A water deficit produces an increase in the ionic concentration of the extracellular compartment. On the contrary, when the body contains an excess of water, the lower ionic concentration of body fluids allows more water to reach the intracellular compartment: drinking is inhibited and the kidneys excrete more water (Ramsay 1998).

Daily water requirement can be defined as the water amount necessary to preserve the water balance of the body. In this ambit, daily water requirements for men should reach about 1,500 ml (Grassi et al. 1983), while children should assume 1.5 ml/kcal/day as recommended value. For a 7-kg baby, the minimum daily intake is 300 ml: 200 ml for *perspiratio insensibilis*, and 100 ml for renal excretion (Castro and Gambarara 2000). During pregnancy and breastfeeding daily water requirements are increased. It has been calculated that a pregnant woman needs a 30 ml/day water increase for amniotic liquid formation and foetus growth. Moreover, during breastfeeding about 650–700 ml/day should be added (SINU 1997).

Water requirement is also related to the level of physical activity. It has been calculated that in an adult of 70 kg performing medium-intensity exercise, daily water balance is 2500 ml (range 1,500–3,000 ml) (Grassi et al. 1983). On the other hand, at rest daily water requirement is 1 ml/kcal in an adult (SINU 1997). Anyway, the human body requires a minimum intake of water in order to be able to sustain life before dehydration occurs. Moreover, young children, pregnant or lactating women, the elderly, the terminally ill and athletes have particular hydration requirements.

Water intake is done for 70–80% through consumption of drinking water and beverages, and for 20–30% through consumption of foods. Diets rich in vegetables and fruit provide significant amounts of the total water intake: the content of water is usually about 90% in fruit, vegetables and milk (EFSA 2010). On the contrary, water loss is mainly due to excretion of water in urine, faeces and sweat; it is higher in physically active people and in hot climates (Malisova et al. 2012).

In 2004, the Institute of Medicine (IOM) set the amount at around 2.7 l of total water a day for women and an average of around 3.7 daily litres for men. This quantity refers to the total daily fluid intake from all sources, defined as '*the amount of water consumed from foods, plain drinking water, and other beverages*' (IOM 2004).

In Italy, the adequate daily intake of water, defined by the 'Recommended Daily Allowances', is referred to gender, chronological age and specific life conditions (pregnancy and breastfeeding) (SINU 1997).

The U.S. Dietary Guidelines 2015–2020 do not recommend a specific daily water or fluid intake, and there is no set upper level for water intake. Anyway, the minimum requirement for water is the amount that equals losses and prevents adverse effects of insufficient water, such as dehydration.[2] In 2010, the European Food Safety Authority (EFSA) used a different approach in developing dietary reference intakes. Water Adequate Intakes (AI) values for various life-stage groups of population were derived from three factors: observed intakes of European population groups, desirable urine osmolality values and desirable total water intake (TWI) volumes per unit of

---

[2]This information can be found at the following web address: https://health.gov/dietaryguidelines/2015/resources/20152020_Dietary_Guidelines.pdf.

dietary energy (Kcal) consumed. With specific relation to water in general situations (moderate environmental temperature and physical activity), EFSA recommends that:

(a) In exclusively breastfed infants during the first half of their first year of life, adequate intakes are 100–190 ml/kg per day, and they are based on water intake from human milk.

(b) For the age period 6–12 months, a total water intake of 800–1,000 ml/day is adequate.

(c) For the second year of life, an adequate total water intake of 1,100–1,200 ml/day is suggested.

(d) For boys and girls with 2–3 years of age, a water intake of 1,300 ml/day is preferred.

(e) For older children (>3 years of age), the adequate total intake increases with age: 1,600 ml/day for boys and girls (4–8 years); 2,100 ml/day for boys (9–13 years); 1,900 ml for girls (9–13 years).

(f) Adolescents of 14 years and older are considered as adults with respect to adequate water intake. Adequate total water intakes should be 2.0 l/day for females and 2.5 l/day for males.

(g) Water losses incurred under extreme conditions of external temperature and physical exercise.

(h) Finally, despite a lower energy requirement, the water requirement in the elderly per unit of dietary energy becomes higher because of a decrease in renal concentrating capacity. For pregnant women, the EFSA proposes the same water intake as in non-pregnant women plus an increase in proportion to the increase in energy intake (300 ml/day), while an additional amount of water (700 ml/day) is recommended for lactating women (in comparison with non-lactating women) (EFSA 2017). However, according to the other scientific societies, the EFSA report stated that a single water intake cannot meet the needs of everyone in a population group because the individual need for water is related to caloric consumption, the concentrating-diluting capacities of the kidneys, and water losses via excretion and secretion.

References

Bedogni G, Borghi A, Battistini N (1996) The assessment of body hydration and water distribution in health and disease. Clin Dietol 21:3–8

Callegaro CC, Moraes RS, Negrão CE, Trombetta IC, Rondon MU, Teixeira MS, Silva SC, Ferlin EL, Krieger EM, Ribeiro JP (2007) Acute water ingestion increases arterial blood pressure in hypertensive and normotensive subjects. J Hum Hypertens 21(7):564–570. https://doi.org/10.1038/sj.jhh.1002188

Castro M, Gambarara M (2000) Nutrizione clinica in Pediatria. McGraw-Hill, Milano

CDC (2016) Get the facts: drinking water and intake. Division of nutrition, physical activity, and obesity. National Center for Chronic Disease Prevention and Health Promotion, Atlanta. https://www.cdc.gov/nutrition/data-statistics/plain-water-the-healthier-choice.html. Accessed 20 Non 2019

Cian C, Koulmann PA, Barraud PA, Raphel C, Jimenez C, Melin B (2000) Influence of variations of body hydration on cognitive performance. J Psychophysiol 14(1):29–36

Del Toma E (1995) Dietoterapia e nutrizione clinica. Il Pensiero Scientifico, Rome

EFSA (2017) Dietary reference values for nutrients. Summary reports. EFSA Supporting publication 2017: e1512. European Food Safety Authority (EFSA), Parma. https://doi.org/10.2903/sp.efsa.2017.e15121. https://www.efsa.europa.eu/sites/default/files/2017_09_DRVs_summary_report.pdf. Accessed 19 Nov 2019

EFSA Panel on Dietetic Products Nutrition and Allergies (NDA) (2010) Scientific Opinion on Dietary Reference Values for water. EFSA J 8(3):1459–1507. https://doi.org/10.2903/j.efsa.2010.14

Fellin A (2003) Quale acqua per la nostra salute? Ed. Tecniche Nuove, Milan

Grandjean AC (2007) Dehydration and cognitive performance. J Am Coll Nutr 26(5 suppl):549S–554S. https://doi.org/10.1080/07315724.2007.10719657

Grassi M, Messina B, Denaro G (1983) L'acqua e le acque minerali nell'alimentazione. Clin Dietol 10:3–25

IOM (2004) Dietary reference intakes for water, potassium, sodium, chloride, and sulfate. Institute of Medicine of the National Academies (IOM). National Academies Press, Washington, D.C.

Kalhoff H (2003) Mild dehydration: a risk factor of broncho-pulmonary disorders? Eur J Clin Nutr 57(S2):S81–S87. https://doi.org/10.1038/sj.ejcn.1601906

Kleiner SM (1999) Water: an essential but overlooked nutrient. J Am Diet Assoc 99(2):200–206. https://doi.org/10.1016/S0002-8223(99)00048-6

Lieberman HR (2007) Hydration and cognition: a critical review and recommendations for future research. J Am Coll Nutr 26(5 Suppl):555S–561S. https://doi.org/10.1080/07315724.2007.10719658

Mac-Mary S, Creidi P, Marsaut D, Courderot-Masuyer C, Cochet V, Gharbi T, Guidicelli-Arranz D, Tondu F, Humbert P (2006) Assessment of effects of an additional dietary natural mineral water uptake on skin hydration in healthy subjects by dynamic barrier function measurements and clinic scoring. Skin Res Technol 12(3):199–205. https://doi.org/10.1111/j.0909-752X.2006.00160.x

Malisova O, Bountziouka V, Panagiotakos DB, Zampelas A, Kapsokefalou M (2012) The water balance questionnaire: design, reliability and validity of a questionnaire to evaluate water balance in the general population. Int J Food Sci Nutr 63(2):138–144. https://doi.org/10.3109/09637486.2011.607799

Manz F, Wentz A (2005) The importance of good hydration for the prevention of chronic diseases. Nutr Rev 63(suppl 1):S2–S5. https://doi.org/10.1111/j.1753-4887.2005.tb00150.x

Marieb EN, Hoehn K (2007) Fluid, electrolyte, and acid-base balance. In: Human anatomy and physiology, 7th edn. Benjamin-Cummings Publishing Company, San Francisco

McCance RA (2002) McCance; Widdowson's the composition of foods, 6th edn. The Royal Society of Chemistry, Cambridge. https://doi.org/10.1039/9781849735551

Murray B (2007) Hydration and physical performance. J Am Coll Nutr 26(5 Suppl):S542–S548

Naumann J, Biehler B, Lüty T, Sadaghiani C (2017) Prevention and therapy of Type 2 diabetes—what is the potential of daily water intake and its mineral nutrients? Nutrients 9(8):914. https://doi.org/10.3390/nu9080914

Neave N, Scholey AB, Emmett JR, Moss M, Kennedy DO, Wesnes KA (2001) Water ingestion improves subjective alertness, but has no effect on cognitive performance in dehydrated healthy young volunteers. Appet 37(3):255–256. https://doi.org/10.1006/appe.2001.0429

Nicolaidis S (1998) Physiology of thirst. In: Arnaud MJ (ed) Hydration throughout life. John Libbey Eurotext, Montrouge

Nissensohn M, López-Ufano M, Castro-Quezada I, Serra-Majem L (2015) Assessment of beverage intake and hydration status. Nutr Hosp 31(3 Suppl):S62–S69. https://doi.org/10.3305/nh.2015.31.sup3.8753

Peronnet F, Mignault D, du Souich P, Vergne S, Le BL, Jimenez L, Rabasa-Lhoret R (2012) Pharmacokinetic analysis of absorption, distribution and disappearance of ingested water labeled with

D(2)O in humans. Eur J Appl Physiol 112(6):2213–2222. https://doi.org/10.1007/s00421-011-2194-7

Pivarnik JM, Palmer RA (1994) Water and electrolyte balance during rest and exercise. In: Wolinsky I, Hickson JF (eds) Nutrition in exercise and sport, 2nd edn. CRC, Boca Raton

Ramsay DJ (1998) Homeostatic control of water balance. In: Arnaud MJ (ed) Hydration throughout life. John Libbey Eurotext, Montrouge

Sawka MN, Cheuvront SN, Carter R (2005) Human water needs. Nutr Rev 63(1):S30–S39. https://doi.org/10.1111/j.1753-4887.2005.tb00152.x

Shirreffs SM, Maughan RJ (1998) Control of blood volume: Long term and short term regulation. In: Arnaud MJ (ed) Hydration throughout life. John Libbey Eurotext, Montrouge

SINU (1997) Livelli di assunzione raccomandati di energia e nutrienti per la popolazione italiana. LARN, Revisione 1996. Società Italiana di Nutrizione Umana (SINU), Rome

Szinnai G, Schachinger H, Arnaud MJ, Linder L, Keller U (2005) Effect of water deprivation on cognitive-motor performance in healthy men and women. Am J Physiol Regul Integr Comp Physiol 289(1):R275–R280. https://doi.org/10.1152/ajpregu.00501.2004

Ticinesi A, Nouvenne A, Borghi L, Meschi T (2017) Water and other fluids in nephrolithiasis: state of the art and future challenges. Crit Rev Food Sci Nutr 57(5):963–974. https://doi.org/10.1080/10408398.2014.964355

U.S. Department of Health and Human Services and U.S. Department of Agriculture (2015) 2015–2020 Dietary guidelines for Americans, 8th edn, December 2015. U.S. Department of Health and Human Services and U.S. Department of Agriculture, Washington, D.C. http://health.gov/dietaryguidelines/2015/guidelines/. Accessed 19 Nov 2019

Wang Z, Deurenberg P, Wang W, Pietrobelli A, Baumgartner RN, Heymsfield SB (1999) Hydration of fat-free body mass: review and critique of a classic body-composition constant. Am J Clin Nutr 69(5):833–841. https://doi.org/10.1093/ajcn/69.5.833

Welch K (2010) Fluid balance. Learn Disabil Pr 13(6):33–38. https://doi.org/10.7748/ldp2010.07.13.6.33.c7890

Chapter 2
The Bottled Water

Abstract The global bottled water market size is expected to reach around USD 350 billion by 2021, following 10% year-on-year growth. The bottled water market is strictly regulated in the United States of America and the European Union. It offers a range of products, including well water, purified water, mineral water, spring water and functional water. Mineral water's 'special' mineral composition might have the properties favourable to health. The leading reason for the explosion in bottled water consumption is people's perception of bottled water as purer and healthier. Furthermore, the bottled water consumption is related to economic, social and demographic factors which might influence attitudes towards bottled water.

Keywords Bottled water · Code of Federal Regulations · Drinking water · European Union · Natural mineral waters · Spring water · Total dissolved solid

Abbreviations

AWWA-RF	American Water Works Association Research Foundation
AFDO	Association of Food and Drug Officials
CFR	Code of Federal Regulations
CGMP	Current Good Manufacturing Practice
EPA	Environmental Protection Agency
EFSA	European Food Safety Authority
EU	European Union
FFDCA	Federal Food, Drug, and Cosmetic Act
FAO	Food and Agricultural Organization
FDA	Food and Drug Administration
HACCP	Hazard Analysis and Critical Control Points
IFEN	Institut Francais de L'Environnement
IBWA	International Bottled Water Association
LDL	Low-density lipoprotein
NIAS	Non-Intentionally Added Substance
PIA	Purified isophthalic acid

M. A. Coniglio et al., *Non-Intentionally Added Substances in PET-Bottled Mineral Water*, Chemistry of Foods, https://doi.org/10.1007/978-3-030-39134-8_2

TDS	Total dissolved solid
U.S.	United States of America
USD	United States Dollar
WHO	World Health Organization

2.1   Worldwide Bottled Water Consumption

A recent report found that a million plastic bottles are purchased every minute world-wide, with that figure likely to increase another 20% by 2021 (Laville and Taylor 2017).

In 2004, the world consumption of bottled water reached 154 billion litres (41 billion gallons), and Americans alone consumed 26 billion litres (Arnold and Larsen 2006). Since then, the demand for bottled water has been increasing, even in places where tap water is safe to drink (Emily and Janet 2006). The United States of America (U.S.) consumption of bottled water in 2008 was estimated to be 2.3 billion litres or 7.3 litres per person (Beverage Marketing 2009). From 2014 to 2017, it grew to over United States Dollars (USD) 200 billion following a 9% yearly growth and it is expected to reach around USD 350 billion by 2021, following 10% year-on-year growth (The Business Research Company 2018).

In 2018, the number of households primarily using bottled water for household consumption worldwide reached nearly 600 million. That year, Europe was one of the most important bottled water producer and consumer regions in the world, followed by North America and Latin America. In 2018, also the bottled water sales in Asian regions nearly doubled, mainly in India, Pakistan and Japan (Rodwan 2018).

2.2   Why People Prefer Bottled Water to Tap Water?

In the general public perception, bottled water is better than tap water mainly in terms of taste, safety and portability (Diduch et al. 2013; Saylor et al. 2011). Dissatisfaction with tap water sensorial properties like, for example, perceived information from taste, odour, colour and turbidity seems to be particularly relevant (Abrahams et al. 2000; Doria 2006; Grondin et al. 1996; IFEN 2000). Anyway, the importance attributed to each of organoleptic features varies according to time and culture (de França Doria 2009).

In particular, in Western countries and in several Canadian regions, water taste is usually more important than odour or appearance (Doria 2006–2010). In fact, in many cases, consumers choose to drink bottled water because they think it tastes better than tap water, which is considered a sign for better quality. Thus, organoleptic and health-related reasons may contribute to bottled water consumption due to concern about tap water risks (AWWA-RF 1993; Levallois et al. 1999).

Anyway, consumers' health consciousness may be largely influenced also by media messages (Ferrier 2001; Olson 1999). Bottles provide indications about the product, which in turn may be related to health, risk and (perceived) sensorial properties (Levallois et al. 1999; Olson 1999).

Many factors may influence perceptions of water quality, and information provided by the mass media are also important (Doria 2009). A number of demographic variables—age, education, income, ethnic group, gender, environmental behaviour—can influence the bottled water usage (Anadu and Harding 2000; Flynn et al. 1994).

High income and young age were found to be directly associated with the risk perception of drinking water. Some studies found that people with a relatively high income and young people (16–25 years) perceive tap water as less safe, with the result that they are the most frequent users of bottled water. On the other hand, education seems to be inversely associated with the risk perception of drinking water (Kotler et al. 2008; Sajjadi et al. 2016).

Another variable influencing consumers' decision is reported to be apparently linked with performance levels of U.S. small water systems when unable to comply with federal quality standards (Anadu and Harding 2000). Therefore, consumers are more likely to use bottled water as a primary source of water when local water is not considered safe (Hu et al. 2011).

2.3 Bottled Water Regulations

According to the European Community legislation (Directive number 98/38), among the waters for human use there are '[...] waters (treated or not), intended for drinking, used for the food and beverages preparation or for other domestic purposes' [...]. Moreover, drinking water must be '[...] clear, odorless, tasteless, colorless and harmless, that is devoid of pathogenic microorganisms and harmful chemicals to humans' (Commission of the European Communities 1998). The U.S. Food and Drug Administration (FDA) describes bottled water as 'Water that is intended for human consumption and that is sealed in bottles or other containers with no added ingredients except that it may contain safe and suitable antimicrobial agents. Fluoride also may be added within the limits set by the FDA' according to the Code of Federal Regulations (CFR) 165.110 (FDA 2019).

Bottled waters are strictly regulated under law. The Codex Alimentarius is a collection of standards, guidelines and codes of practice adopted by the Codex Alimentarius Commission, which is a joint initiative of the Food and Agricultural Organization of United Nations (FAO) and the World Health Organization (WHO). Codex Standard for Natural Mineral Waters—Codex Standard 108-1981—applies to all packaged drinking natural mineral waters (Codex Alimentarius Commission 1985a). It does not apply to natural mineral waters sold or used for other purposes.

According to the Codex, natural mineral waters have to be bottled (packed) in hermetically sealed retail containers with the aim of preventing possible adulteration or contamination. Naturally, compliance to microbiological criteria as defined in Annex I of the Code of Hygienic Practice for Collecting, Processing and Marketing of Natural Mineral Waters (Codex Alimentarius Commission 1985b) has to be assured. Furthermore, like all the other packaged foods and beverage products, bottled waters have extensive labelling requirements, including a statement of the type of water that is in the container. The General Standard for the Labelling of Prepackaged Foods (Codex Alimentarius Commission 1985c) establishes that

(1) The name of the product shall be 'natural mineral water', with one additional descriptive term.
(2) The location of the source and the name of the source shall be declared (important conditions for 'spring' waters).
(3) The analytical composition justifying certain and claimed features to bottled waters shall be declared in the labelling.

Finally, treatments allowed by the Codex Standard for Natural Mineral Waters for the elimination of metallic unstable elements (examples: sulphur, iron, etc.) include decantation, filtration and possible aeration (as a synergic measure for decantation and filtration). If a natural mineral water has been submitted to one of the mentioned treatments, the result of the treatment shall be declared on the label.

2.3.1   EU Legislation on Bottled Waters

In Europe, specific legislation applies to the three different categories of bottled water listed in Fig. 2.1.

The Drinking Water Directive (Council Directive 98/83/EC) concerns the quality of water intended for human consumption. Its purpose is to set up the essential quality standards at the European Union (EU) level, and to protect human health from adverse effects of any contamination of water intended for human consumption. The Directive also applies to drinking water in bottles, with the exception of natural mineral waters and waters which are medicinal products. The Directive requires a total of 48 microbiological, chemical and indicator parameters must be monitored and tested regularly. When translating the Drinking Water Directive into their own national legislation, EU Member States can introduce additional requirements or different control levels valuable within their territory or set higher standards, provided that standards for protection of human health are not lowered. Nonetheless, Member States are not allowed to set lower standards because standard levels should be equal within the whole EU. The Directive also requires providing regular information to consumers.

The Directive 2009/54/EC of 18 June 2009 *'On the exploitation and marketing of natural mineral waters'* prescribes specific rules for marketing this type of water. According to Article 1(1) and (2) and Article 2 of the Directive 2009/54/EC, waters may only be marketed as 'natural mineral waters' in the European Union if they fulfill the following conditions:

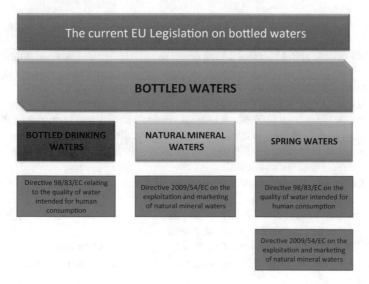

Fig. 2.1 The current EU Legislation on bottled waters

(i) *'Where waters are extracted from the ground of a Member State, the responsible authority of that Member State has recognised the waters as natural mineral waters in accordance with Directive 2009/54/EC'*

(ii) *'Where waters are extracted from the ground of a third country, the responsible authority of a Member State has recognised the waters as natural mineral waters in accordance with Directive 2009/54/EC'.*

Moreover, the following requirements have to be considered, among all mandatory points (European Parliament and Council 2009):

(a) Natural mineral waters may not undergo any (anthropic) treatment apart, for instance, from the separation of certain unstable elements.

(b) Natural mineral waters have to be guaranteed free from parasites, coliform bacteria and other dangerous (for human health) components at the original source.

(c) Used containers for natural mineral waters have to be hermetically sealed with the aim of avoiding contamination.

In addition, according to the Directive, in order to protect the health of consumers as well as to prevent consumers from being misled '[…] *care should be taken to ensure that natural mineral waters retain at the marketing stage those characteristics which enabled them to be recognised as such. Therefore, the containers used for packaging them should have suitable closures'*. Furthermore, *'any containers used for packaging natural mineral waters shall be fitted with closures designed to avoid any possibility of adulteration or contamination'* (Article 6).

The following EU Regulations and Directives on Food Safety and Food Hygiene may be also applied to bottled waters:

- Directive 13/2000 of the European Parliament and the Council relating to the Labelling, Presentation and Advertising of Food Stuffs.
- Regulation (EC) No 178/2002 laying down the general principles and requirements of food law, establishing the European Food Safety Authority and laying down procedures in matters of food safety.
- Regulation (EC) No 852/2004 on the hygiene of foodstuffs.
- Regulation (EC) No 882/2004 on official controls performed to ensure verification of compliance with feed and food law, animal health and animal welfare.
- Commission Directive 19/2004/EC on plastic materials and articles intended to come into contact with foodstuffs.
- Regulation 1924/2006 of the European Parliament and the Council on Nutrition and Health Claims made on Foods.
- Commission Regulation (EC) No 282/2008 on recycled plastic materials and articles intended to come into contact with foodstuffs.
- Guidelines for the development of EU Guides to Good Practice for hygiene or for the application of the 'Hazard Analysis and Critical Control Points' (HACCP) principles.

Finally, EU legislation authorises the use of a number of indications (examples: '*stimulates digestion*'; '*diuretic*') related to the composition and related health benefits for these products. Also, EU Member States can establish peculiar criteria in this ambit.

2.3.2 U.S. Legislation on Bottled Waters

In the U.S., bottled drinking water is regulated as a food by the Food and Drug Administration (FDA) by means of the Federal Food, Drug, and Cosmetic Act (FFDCA) concerning foods introduced or delivered for introduction in the domestic market. The manufacturer of bottled waters is fully responsible for assuring that bottled waters introduced in interstate commerce comply with all FFDCA requests and FDA Regulations in this ambit.

FDA has established specific regulations for bottled water in Title 21 of the Code of Federal Regulations (21 CFR), including (section 165.110[a]—standards of identity regulations that define different types of bottled water, such as 'artesian water', '*artesian well water*', 'groundwater', 'mineral water', 'purified water', '*sparkling bottled water*', and '*spring water*' (FDA 2019). Bottled water labelled with any of these terms must meet the appropriate definitions under the standard of identity (21 CFR part 101). For example, a bottle labelled as containing 'mineral water' must meet the following criteria (FDA 2019):

(1) The water must contain no less than 250 parts per million (ppm) total dissolved solids (TDS).
(2) It must contain no added minerals.

(3) It must come from a geologically and physically protected underground water source.

FDA has also established Current Good Manufacturing Practice (CGMP) Regulations for the processing and bottling of bottled drinking water (21 CFR part 129). In summary, bottled water has to be the result of adequate processing, bottling, storing and transportation steps when speaking of human health, food safety and possible contamination risks for the original water source. Consequently, adequate quality controls are needed to assure the bacteriological and chemical safety of the water, sampling and testing of source water and the final product (21 CFR section 165.110[b]).

Bottlers/manufacturers are responsible for ensuring that their bottled water can pass tests used by FDA. As a result, FDA is responsible for periodical inspections at bottled water plants and the collection of bottled water samples. Investigated parameters in FDA laboratories are generally related to microbiological, radiological or chemical contamination; in addition, selected contaminants may be investigated depending on the reason for the sampling.

In the U.S., the Environmental Protection Agency (EPA) regulates tap water with extensive norms ruling the production, distribution and quality of drinking water, including contaminant levels. Consequently, FDA is required to (FDA 2001)

(a) Establish quality standard regulations for bottled water in response to developments at EPA.
(b) Establish a standard of quality regulations for contaminant levels in bottled waters if EPA defines new maximum contaminant levels or treatment techniques for contaminants in public drinking.
(c) Consider the inclusion of possible disinfectants in bottled waters by means of related regulations. The adoption of EPA's allowable levels of three disinfectants such as chlorine dioxide and four disinfection by-products including haloacetic acids and total trihalomethanes is one of these examples concerning quality standard for bottled water.

Finally, professional trade organisations are allowed to regulate the water industry in the U.S. beyond government norms. The International Bottled Water Association (IBWA) has defined industrial standards with the aim of raising the quality of bottled waters if compared with FDA or EPA requirements. Of course, the IBWA lacks the same enforcement authority of a government agency, but it may be able to bring government agencies' attention to members who are in violation of safety standards.

2.4   Mineral Water and Natural Mineral Water

Mineral water means water containing not less than 250 parts per million (ppm) of total dissolved solids (TDS), and originating from a geologically and physically protected underground water source. Natural mineral water is not derived from a municipal system or public water supply and is unmodified except for limited treatment (e.g. filtration, ozonation or equivalent disinfection process).

2.5   Types of Bottled Waters

According to the U.S. Food and Drug Administration (FDA), bottled water is *'intended for human consumption and sealed in bottles or other containers with no added ingredients, except that it may contain a safe and suitable antimicrobial agent […]'*. However, it should be also mentioned that *'drinking water'* is also intended to be treated by means of filtration and ozonation processes, although similar disinfection systems can be considered in his ambit (FDA 2019).

In this situation, *'bottled water'* means all possible *'bottled water products'*; on the contrary, the apparent synonym *'drinking water'* does not mean or include all possible *'bottled water products'*. For this and other reasons, the classification of bottled water products should be briefly provided and discussed when speaking of FDA requirements. Six main typologies may be discriminated in the ambit of bottled water products (FDA 2019):

(a)   Natural Mineral Water (Sect. 2.5.1),
(b)   Spring Water (Sect. 2.5.2),
(c)   Purified Water (Sect. 2.5.3),
(d)   Artesian Water (Sect. 2.5.4),
(e)   Functional Water (Sect. 2.5.5) and
(f)   Fluoridated Water (Sect. 2.5.6).

2.5.1   Mineral Water

Natural mineral waters are *'originated from an aquifer or underground reservoir, spring from one or more natural or bore sources and have specific hygienic features and, eventually, healthy properties'* (Directive 2009/54/EC). On these bases, natural mineral waters have to be

(1)   Pure and safe as the original water at source.
(2)   Free from chemical treatments (and related residual components).
(3)   Tasteful because of a specific mineral quantitative and qualitative composition. This specificity should be distinctive and peculiar of the original source.

FDA considers that mineral natural water has to be discriminated from other waters if this product maintains constantly its mineral substances and traceable elements without differences from the origin to bottling, and if TDS $\geq$ 250 ppm at least, as suggested by the Association of Food and Drug Officials (AFDO). This requirement does not force manufacturers to use other possible adjectives ('ground', well' or 'spring') in the same ambit.

However, consumers might argue that 'mineral waters' can contain a significant quantity of mineral substances. As a result, the following statement (or similar phrases): '*Not a significant source of ____*' could be added to labels concerning mineral waters. Naturally, such a statement should suggest that the reported element(s) are missing or negligible in the mineral water when speaking of mineral concentration. Manufacturers are also allowed to state and declare TDS levels on labels.

In the European Union (2009/54/EC Directive, Annex 3; Fig. 2.2), mineral waters may be classified on the basis of the declared (on labels) salt mineral composition as mg/litre (on the total amount of mineral residuals resulting from initial evaporation of 1 l of water at 100 °C and final drying at 180 °C). 'Minimally mineralised waters' have a TDS < 50 mg/l; 'oligomineral waters' remain between 50 and 500 mg/l, while 'medium mineralised' and 'rich in minerals' waters: exhibit TDS between 500 and 1,500 mg/l, and >1,500 mg/l, respectively (Fig. 2.2).

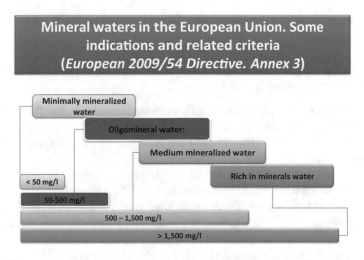

Fig. 2.2 The current situation of mineral waters in the European Union. Some indications and related criteria according to the European 2009/54 Directive. Annex 3

Natural Mineral Waters. A simplified classification...

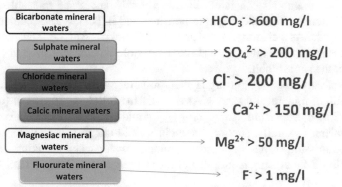

Fig. 2.3 A simplified representation of mineral waters in the current market. This picture has been realised by Carmelo Parisi, currently a student at the Liceo Scientifico Stanislao Cannizzaro, Palermo, Italy

2.5.1.1   Natural Mineral Waters. Mineral Composition and Health Effects

With relation to natural mineral waters, the following European classification has to be also taken into account (Fig. 2.3):

- Bicarbonate mineral waters. The bicarbonate ion has to be higher than 600 mg/l.
- Sulphate mineral waters. The sulphate ion is required to be >200 mg/l.
- Chloride mineral waters. The chloride ion is required to be >200 mg/l, similarly to sulphate ion.
- Calcic mineral waters. The calcium ion has to be higher than 150 mg/l.
- Magnesiac mineral waters. Magnesium ion has to be >50 mg/l.
- Fluorurated mineral waters. The fluoride ion has to be >1 mg/l.
- Waters with sodium. In this ambit, and similarly to sulphate and chloride ions, sodium content is required to be >200 mg/l.
- Acid waters. Free carbon dioxide (CO_2) has to be higher than 250 mg/l.

Depending on the composition of the water, mineral waters may significantly help the human being when speaking of recommended daily intakes, because of the exogenous nature of mineral compounds. Living organisms cannot be able to synthesise minerals (European Food Safety Authority 2017).

Calcium

Calcium, as metallic cation (Ca^{2+}), is needed for human beings. Recommended daily intakes are: adults, up to 1,000 mg; adolescents (11–17 years), up to 1,150 mg.

Bicarbonate

Bicarbonate (hydrogen carbonate, HCO_3^-) may be generated in the body, but generally in insufficient quantities. There is no recommended daily allowance.

Magnesium

Magnesium, as metallic cation (Mg^{2+}), is needed for human beings. Recommended daily intakes are: adults, 300–400 mg; adolescents (11–17 years), 250–300 mg.

Chloride

Chloride, as halogen anion (Cl^-), can be obtained only by food ingestion. Recommended daily intakes are 2,300 mg/l in general.

Sodium

Sodium, as mineral cation (Na^+), is needed for human beings. Recommended daily intakes are generally 1,500 mg/l. Interestingly, in the EU, labels can show two claims: *'contains sodium'* (if sodium $\geq$ 200 mg/l) or *'suitable for a low-sodium diet'* (if sodium is lower than 20 mg/l).

Sulphates

Sulphate, as anion (SO_4^{2-}), cannot be produced in the body. Interestingly, there is no recommended daily allowance in this ambit.

2.5.1.2   Natural Mineral Waters. General Features

Natural mineral waters are commonly used to prevent or improve various diseases.

With reference to cations, calcic mineral waters can effectively augment serum calcium. According to the European Food Safety Agency (EFSA), mineral waters with high calcium contents may be consumed as 'functional foods' (EFSA 2016). In addition, parathyroid hormone secretion is reported to be inhibited if calcium is abundant in the body, with related positive effects on bones and also non-skeletal systems, including also post-menopausal women (Aptel et al. 1999; Heaney 2006; Meunier et al. 2005). Magnesiac waters can have positive effects in women entering pre-menstrual and climaterium/post-menopausal periods; in general, risk of coronary heart disease appears to be reduced (Jiang et al. 2016; Petraccia et al. 2006). Sodium-rich mineral waters can help against chronic dyspepsia (Bortolotti et al. 1999), cardiovascular diseases (Schoppen et al. 2004) and insulin-related problems (Schoppen et al. 2005).

When speaking of anions, chloride mineral waters (chloride is the common partner for sodium, calcium and magnesium) may have good effects against chronic gastric diseases (Chaban et al. 1990; Petraccia et al. 2006). Bicarbonate-rich mineral waters are reported to have positive results on the health level when speaking of low-density lipoprotein (LDL)-cholesterol control, and renal, bone-related, cardiovascular and glycemic dysfunctions (Bertoni et al. 2002; Burckhardt 2008; Murakami et al. 2015; Schoppen et al. 2004–2005; Toxqui and Vaquero 2016).

Different associations between anions and cations have been also reported by the health viewpoint when speaking of magnesium and sodium sulphate mineral waters (Dupont et al. 2014), sulphate-bicarbonate-calcium-magnesiac (Mennuni et al. 2014), calcic and magnesiac sulphate-sulphurous (Toussaint et al. 1986–1988) and waters containing sodium and bicarbonate with concomitant low TDS values (Schorr et al. 1996). In addition, fluorurate waters may be helpful against pyorrhoea and dental decays, while acid waters are reported to exert positive effects against constipation, dyspepsia (Cuomo et al. 2002) and in therapies improving pharyngeal swallowing (Morishita et al. 2014).

2.5.2 Spring Water

'Spring water' is a difficult term with relation to waters, because of the implicit meaning of 'pure' and 'natural' water (anthropic activities would be completely absent). Actually, spring water is '*water derived from an underground formation from which water flows naturally to the surface of the earth, or would flow naturally to the surface of the earth if not for its collection below the earth's surface*' for the FDA. In addition, FDA considers water from a borehole drilled next to the spring as 'spring water' (this method is also considered as a safe and reliable system in the U.S. and the EU), provided that four specific conditions are fully considered (FDA 2019; Mascha 2006). In general, springs are usually classified by the volume of water they discharge. Anyway, 'spring waters' have to exhibit the same properties of waters flowing naturally (without anthropic activities), and this requirement includes also safe levels for many contaminants.

Should water properties be different because of external (anthropic) causes, the 'spring' attribution could not be longer used. In this ambit, it should be assumed that external activities have altered water properties, with a needed period for the restoration of natural equilibrium. Additionally, the position of the spring has to be mandatorily identified.

2.5.3   Purified Water

'Purified' water is modified in some ways, by means of purification treatments such as filtration, distillation, reverse osmosis, etc. (FDA 2019). In this situation, waters may be also named according to the specified and labelled treatment(s). Anyway, TDS must be no more than 10 ppm and pH between 5 and 7. Microbiologically, these waters are pure although sterility cannot be guaranteed (Keyashian 2014). Ions such as chloride, calcium, sulphate and other chemicals (ammonia, carbon dioxide, heavy metals, etc.) have to be absent. For these reasons, manufacturers/bottlers may also mention specific information on labels, especially with relation to peculiar consumer groups with specific needs (immunosuppressed people, etc.).

2.5.4   Artesian Water

'Artesian' or 'artesian well' water is labelled in this way if it comes from underground sources tapping layers of porous rock, sand and earth. For these reasons, also, water may naturally flow because of inner pressure, although surfaces may be not always reached (FDA 2019). In certain situations (artesian well waters), manufactures would be requested to demonstrate that well waters come from real artesian wells. Anyway, external (anthropic) activities for collection are allowed, although physical properties, chemical composition and quality may be altered. In these situations, the manufacturer is requested to demonstrate that no alteration is possible.

2.5.5   Functional Water

Functional water is water that has either had herbs or fruit extracts, or antioxidants, or vitamins added to it, or has been electromechanically modified at the molecular level to give it a new chemical structure and to endow it with new functions. Thus, functional water is a beverage that is supposed to provide additional functions other than hydration. The most popular functional waters are added with lime, strawberry, *Aloe vera*, etc., to give the consumer a vitamin boost. Other functional waters contain higher magnesium, chloride or sodium levels.

The largest markets of functional waters are Asia Pacific, Western Europe and North America. The main reason why so many consumers use functional water is, of course, health concerns. In particular, consumers who want to lose weight or stay slim and fit will prefer low or zero-calorie beverages. Anyway, functional water consumers are not only health conscious, but they are also increasingly environmentally aware. For this reason, functional water beverages are being sold more and more in glass containers or in special biodegradable packaging.

2.5.6 Fluoridated Water

Fluoridated water can be defined as bottled water containing naturally occurring or added fluoride. The label should specify whether fluoride is naturally occurring or added. Anyway, any water that is called 'fluoridated water' should contain not less than 0.8 mg per litre of fluoride ion. Fluoridated waters may be indicated for children because they can reduce the caries process, as well as the incidence of decay. It is now recognised that the main effect of water fluoridation is local and post-eruptive. Fluoridated waters may also promote bone mineralisation, and fluoride supplements are generally prescribed for breast-fed infants because those infants frequently consume little or no water (Cury et al. 2019).

Anyway, the use of fluoride in water to control caries has created a controversy due to data associating water fluoridation as the cause of some systemic diseases. In fact, high fluoridated mineral water consumption may have some toxic effects: from dental fluorosis to skeletal fluorosis, if fluoride intake is above 10 mg/l (Den Besten et al. 2011). For these reasons, the EFSA established fluoride upper limit of exposure to 1.5 mg/l/die (EFSA 2005). This value limit is confirmed also by the World Health Organization (World Health Organization 2011).

In addition to these three major categories of bottled waters, the International Bottled Water Association (IBWA) considers the following four other categories (IBWA 2009):

- Well water. This is water from a hole bored or drilled into the ground, which taps into an aquifer.
- Sparkling bottled water. After treatment and possible replacement of carbon dioxide, this water contains the same amount of carbon dioxide from the source that it had at emergence from the source.

References

Abrahams N, Hubbell B, Jordan J (2000) Joint production and averting expenditure measures of willingness to pay: do water expenditures really measure avoidance costs? Am J Agric Econ 82(2):427–437. https://doi.org/10.1111/0002-9092.00036

Anadu EC, Harding AK (2000) Risk perception and bottled water use. J Amer Water Works Assoc 92(11):82–92. https://doi.org/10.1002/j.1551-8833.2000.tb09051.x

Aptel I, Cance-Rouzaud A, Grandjean H (1999) Association between calcium ingested from drinking water and femoral bone density in elderly women: evidence from the EPIDOS cohort. J Bone Miner Res 14(5):829–833. https://doi.org/10.1359/jbmr.1999.14.5.829

Arnold E, Larsen J (2006) Bottled water: pouring resources down the drain. Earth Policy Institute, Rutgers University, New Brunswick. http://www.earth-policy.org/plan_b_updates/2006/update51. Accessed 20 Nov 2019

Bertoni M, Olivieri F, Manghetti M, Boccolini E, Bellomini MG, Blandizzi C, Bonino F, Del Tacca M (2002) Effects of a bicarbonate-alkaline mineral water on gastric functions and functional dyspepsia: a preclinical and clinical study. Pharmacol Res 46(6):525–531. https://doi.org/10.1016/S1043661802002323

Beverage Marketing (2009) Smaller categories still saw growth as the U.S. liquid refreshment beverage market shrank by 2.0% in 2008. Beverage Marketing, New York, 31 August 2009

Bortolotti M, Turba E, Mari C, Lopilato C, Porrazzo G, Scalabrino A, Miglioli M (1999) Changes caused by mineral water on gastrointestinal motility in patients with chronic idiopathic dyspepsia. Minerva Med 90(5–6):187–194

Burckhardt P (2008) The effect of the alkali load of mineral water on bone metabolism: interventional studies. J Nutr 138(2):435S–437S. https://doi.org/10.1093/jn/138.2.435S

Chaban AG, Lysiuk AD, Chernobrovyi VN, Kuchuk AP (1990) The therapeutic efficacy of a sodium chloride mineral water in chronic gastritis patients with secretory insufficiency. Vopr Kurortol Fizioter Lech Fiz Kult 5:17–19

Codex Alimentarius Commission (1985a) CODEX STAN 108-1981. Codex standard for natural mineral waters. Food and Agriculture Organization of the United Nations (FAO) and World Health Organization (WHO), Rome. http://www.fao.org/input/download/standards/223/CXS_108e.pdf. Accessed 20 Nov 2019

Codex Alimentarius Commission (1985b) CAC/RCP 33-1985—Codex Alimentarius. Code of hygienic practice for collecting, processing and marketing of natural mineral waters. Joint FAO/WHO Food Standards Programme, Food and Agriculture Organization of the United Nations (FAO) and World Health Organization (WHO), Rome

Codex Alimentarius Commission (1985c) CODEX STAN 1-1985. General standard for the labelling of prepackaged foods. Food and Agriculture Organization of the United Nations (FAO) and World Health Organization (WHO), Rome. http://www.fao.org/3/Y2770E/y2770e02.htm. Accessed 20 Nov 2019

Commission of the European Communities (1998) Commission Directive 98/38/EC of 3 June 1998 adapting to technical progress Council Directive 74/151/EEC on certain components and characteristics of wheeled agricultural or forestry tractors. Off J Eur Comm L170:13–14

Cuomo R, Grasso R, Sarnelli G, Capuano G, Nicolai E, Nardoni G, Pomponi D, Budillon G, Ierardi E (2002) Effects of carbonated water on functional dyspepsia and constipation. Eur J Gastroenterol Hepatol 14(9):991–999

Cury JA, Ricomini-Filho AP, Berti FLP, Tabchoury CP (2019) Systemic effects (risks) of water fluoridation. Braz Dent J 30(5):421–428. https://doi.org/10.1590/0103-6440201903124

de França Doria M (2009) Factors influencing public perception of drinking water quality. Water Policy 12(1):1–19. https://doi.org/10.2166/wp.2009.051

Den Besten P, Li W (2011) Chronic fluoride toxicity: dental fluorosis. Monogr Oral Sci 22:81–96. https://doi.org/10.1159/000327028

Diduch M, Polkowska Z, Namieśnik J (2013) Factors affecting the quality of bottled water. J Expo Sci Environ Epidemiol 23(2):111–119. https://doi.org/10.1038/jes.2012.101

Doria MD (2006) Bottled water versus tap water: understanding consumers' preferences. J Water Health 4(2):271–276. https://doi.org/10.2166/wh.2006.0023

Dupont C, Campagne A, Constant F (2014) Efficacy and safety of a magnesium sulfate-rich natural mineral water for patients with functional constipation. Clin Gastroenterol Hepatol 12(8):1280–1287. https://doi.org/10.1016/j.cgh.2013.12.005

EFSA (2017) Dietary Reference Values for nutrients. Summary reports. EFSA Supporting publication 2017:e1512. European Food Safety Authority (EFSA), Parma. https://doi.org/10.2903/sp.efsa.2017.e15121. https://www.efsa.europa.eu/sites/default/files/2017_09_DRVs_summary_report.pdf. Accessed 19 Nov 2019

EFSA NDA Panel, Turck D, Bresson JL, Burlingame B, Dean T, Fairweather-Tait S, Heinonen M, Hirsch-Ernst KI, Mangelsdorf I, McArdle H, Naska A, Neuhäuser-Berthold M, Nowicka G, Pentieva K, Sanz Y, Sjödin A, Stern M, Tomé D, Van Loveren H, Vinceti M, Willatts P, Martin A, Strain S, Siani A (2016) Scientific opinion on calcium and contribution to the normal development of bones:evaluation of a health claim pursuant to Article 14 of Regulation (EC) No 1924/2006. EFSA J 14(10):4587–4596. https://doi.org/10.2903/j.efsa.2016.4587

EFSA Panel on Contaminants (2005) Opinion of the Scientific Panel on Contaminants in the food chain on a request of the Commission related to concentration limits for boron and fluoride in natural mineral waters adopted on 22 June 2005. EFSA J 237:1–8. https://doi.org/10.2903/j.efsa.2005.237

Emily A, Janet L (2006) Plan B updates. Earth Policy Institute, Rutgers University, New Brunswick, NJ. Available http://www.earth-policy.org/index.php?/plan_b_updates/2006/update51. Accessed 27 Dec 2019

European Council (1998) Council Directive 98/83/EC of 3 November 1998 on the quality of water intended for human consumption. Off J Eur Comm L330:0032–0054

European Parlament and Council (2009) Directive 2009/54/EC of the European Parliament and of the Council of 18 June 2009 on the exploitation and marketing of natural mineral waters. Off J Eur Union L164:45–58

FDA (2001) 66 FR 35373—beverages: bottled water; technical amendment; confirmation of effective date. Food and Drugs Administration (FDA), Department of Health and Human Services, Washington, DC

FDA (2019) Code of Federal Regulations. Title 21. Food for human consumption. Food and Drugs Administration (FDA), Department of Health and Human Services, Wagshington. DC

Ferrier C (2001) Bottled water: understanding a social phenomenon. Ambio 30(2):118–119. https://doi.org/10.1579/0044-7447-30.2.118

Flynn J, Slovic P, Mertz CK (1994) Gender, race, and perception of environmental health risks. Risk Anal 14(6):1101–1108. https://doi.org/10.1111/j.1539-6924.1994.tb00082.x

Grondin J, Levallois P, Moret S, Gingras S (1996) The influence of demographics, risk perception, knowledge, and organoleptics on water consumption patterns. In: Proceedings of the American Water Works Association (AWWA) annual conference: management and regulations, Denver, 1996

Heaney P (2006) Absorbability and utility of calcium in mineral waters. Am J Clin Nutr 84(2):371–374. https://doi.org/10.1093/ajcn/84.2.371

Hu Z, Morton L, Mahler RL (2011) Bottled water: United States consumers and their perceptions of water quality. Int J Environ Res Public Health 8(2):565–578. https://doi.org/10.3390/ijerph8020565

Hurd R (1993) Consumer attitude survey on water quality issues. American Water Works Association Research Foundation (AWWA-RF), Denver

IBWA (2009) Bottled water code practice. International Bottled Water Association (IBWA), Alexandria. Available https://www.bottledwater.org/files/IBWA%20Code%20of%20Practice%20Updated%202009%20Final_3.pdf. Accessed 21 Nov 2019

IFEN (2000) La préoccupation des Français pour la qualité de l'eau. Les Données de 'Environnement 57:1–4. Institut Francais de L'Environnement (IFEN), Orléans. http://www.side.developpement-durable.gouv.fr/EXPLOITATION/DEFAULT/Infodoc/ged/viewportalpublished.ashx?eid=IFD_FICJOINT_0001678&search=. Accessed 20 Nov 2019

Jiang L, He P, Chen J, Liu Y, Liu D, Qin G, Tan N (2016) Magnesium levels in drinking water and coronary heart disease mortality risk: a meta-analysis. Nutrients 8(1, pii):E5. https://doi.org/10.3390/nu8010005

Keyashian M (2014) Water systems for pharmaceutical facilities. In: Vogel HC, Todaro CM (eds) Fermentation and biochemical engineering handbook, 3rd edn. William Andrew, Waltham and Kidlington. https://doi.org/10.1016/b978-1-4557-2553-3.00017-9

Kotler P, Armstrong G, Wong V, Saunders J (2008) Principles of Marketing, 5th edn. Pearson Education Limited, Harlow

Laville S, Taylor M (2017) A million bottles a minute: world's plastic binge 'as dangerous as climate change'. Guardian News & Media Limited London. https://www.theguardian.com/environment/2017/jun/28/a-million-a-minute-worlds-plastic-bottle-binge-as-dangerous-as-climate-change. Accessed 20 Nov 2019

Levallois P, Grondin J, Gingras S (1999) Evaluation of consumer attitudes on taste and tap water alternatives in Quebec. Wat Sci Technol 40(6):135–139. https://doi.org/10.1016/S0273-1223(99)00549-1

Mascha M (2006) Fine waters: A connoisseur's guide to the world's most distinctive bottled waters. Quirk Books, Philadelphia

Mennuni G, Petraccia L, Fontana M, Nocchi S, Stortini E, Romoli M, Esposito E, Priori F, Grassi M, Geraci A, Serio A, Fraioli A (2014) The therapeutic activity of sulphate-bicarbonate-calcium-magnesiac mineral water in the functional disorders of the biliary tract. Clin Ter 165(5):e346–e352. https://doi.org/10.7417/CT.2014.1761

Meunier PJ, Jenvrin C, Munoz F, de la Gueronnière V, Garnero P, Menz M (2005) Consumption of a high calcium mineral water lowers biochemical indices of bone remodeling in postmenopausal women with low calcium intake. Osteoporos Int 16(10):1203–1209. https://doi.org/10.1007/s00198-004-1828-6

Morishita M, Mori S, Yamagami S, Mizutani M (2014) Effect of carbonated beverages on pharyngeal swallowing in young individuals and elderly inpatients. Dysphagia 29(2):213–222. https://doi.org/10.1007/s00455-013-9493-6

Murakami S, Goto Y, Ito K, Hayasaka S, Kurihara S, Soga T, Tomita M, Fukuda S (2015) The consumption of bicarbonate-rich mineral water improves glycemic control. Evid Based Complement Altern Med 2015: Article ID 824395. https://doi.org/10.1155/2015/824395

Olson ED (1999) Bottled water: pure drink or pure hype? Natural Resources Defense Council (NRDC), Publications Department, New York

Petraccia L, Liberati G, Masciullo SG, Grassi M, Fraioli A (2006) Water, mineral waters and health. Clin Nutr 25(3):377–385. https://doi.org/10.1016/j.clnu.2005.10.002

Rodwan JG Jr (2018) Bottled water 2017. Bottled Water Reporter Jul/Aug 2018:12–20.www.bottledwater.com. https://www.bottledwater.org/public/BMC2017_BWR_StatsArticle.pdf. Accessed 20 Nov 2019

Sajjadi SA, Alipour V, Matlabi M, Biglari H (2016) Consumer perception and preference of drinking water sources. Electron Physician 8(11):3228–3233. https://doi.org/10.19082/3228

Saylor A, Prokopy LS, Amberg S (2011) What's wrong with the tap? Examining perceptions of tap water and bottled water at Purdue University. Environ Manag 48(3):588–601. https://doi.org/10.1007/s00267-011-9692-6

Schoppen S, Pérez-Granados AM, Carbajal A, Oubiña P, Sánchez-Muniz FJ, Gómez-Gerique JA, Vaquero MP (2004) A sodium-rich carbonated mineral water reduces cardiovascular risk in postmenopausal women. J Nutr 134(5):1058–1063. https://doi.org/10.1093/jn/134.5.1058

Schoppen S, Pérez-Granados AM, Carbajal A, Sarriá B, Sánchez-Muniz FJ, Gómez-Gerique JA, Pilar Vaquero M (2005) Sodium bicarbonated mineral water decreases postprandial lipaemia in postmenopausal women compared to a low mineral water. Br J Nutr 94(4):582–587. https://doi.org/10.1079/BJN20051515

Schorr U, Distler A, Sharma AM (1996) Effect of sodium chloride- and sodium bicarbonate-rich mineral water on blood pressure and metabolic parameters in elderly normotensive individuals: a randomized double-blind crossover trial. J Hypertens 14(1):131–135

The Business Research Company (2018) The global bottled water market: expert insights & statistics. MarketResearch.com, Rockville. https://blog.marketresearch.com/the-global-bottled-water-market-expert-insights-statistics. Accessed 20 Nov 2019

Toussaint C, Peuchant E, Nguyen BC, Jensen R, Canelas J (1986) Influence of calcic and magnesic sulphurous thermal water on the metabolism of lipoproeins in the rat. Arch Int Physiol Biochim 94:65–76

Toxqui L, Vaquero MP (2016) An intervention with mineral water decreases cardiometabolic risk biomarkers. A crossover, randomised, controlled trial with two mineral waters in moderately hypercholesterolaemic adults. Nutrients 8(7):E400. https://doi.org/10.3390/nu8070400

WHO (2011) Fluoride. Drinking water guidelines, 4th edn. Chemical Fact Sheets. World Health Organization (WHO), Geneva

Chapter 3
Polyethylene Terephthalate

Abstract Polyethylene terephthalate is the most widely used packaging material worldwide for bottled water and other non-alcoholic refreshment beverages. However, in recent years, concerns are rising about the safety of polyethylene terephthalate food packaging due to the possible migration of chemical compounds from polyethylene terephthalate bottles into the water contained in it which may pose health risk to consumers. In Europe, the EU Regulation No 10/2011 on food packaging materials establishes a positive list of the compounds authorised for use in plastic materials intended to come into contact with food and provides migration limits for several molecules. In the United States, the Food and Drug Administration regulates food contact substances, materials and articles. All existing regulations are published in the United States Code of Federal Regulation.

Keywords Food contact material · Monoethylene glycol · Non-intentionally added substance · Overall migration limit · Polyethylene terephthalate · Specific migration limit · Terephthalic acid

Abbreviations

MXD-6	1,3-benzenedimethanamine
CHDM	1,4-cyclohexanedimethanol
DEHP	Bis(2-ethylhexyl) phthalate
BHET	Bis-hydroxyethyl-terephthalate
BHT	Butylated hydroxytoluene
CFR	Code of Federal Regulations
DoC	Declaration of Compliance
DBP	Dibutyl phthalate
DEP	Diethyl phthalate
DiisoBP	Diisobutyl phthalate
DMP	Dimethyl phthalate
DMT	Dimethyl terephthalate
EA	Environmental Assessment

M. A. Coniglio et al., *Non-Intentionally Added Substances in PET-Bottled Mineral Water*,
Chemistry of Foods, https://doi.org/10.1007/978-3-030-39134-8_3

EIS	Environmental Impact Statement
EPA	Environmental Protection Agency
EG	Ethylene glycol
EFSA	European Food Safety Authority
EU	European Union
FONSI	Finding of No Significant Impact
FDA	Food and Drug Administration
FCM	Food Contact Material
FCS	Food Contact Substance
IBWA	International Bottled Water Association
IPA	Isophthalic acid
MTR	Melt to Resin
MEG	Monoethylene glycol
NIAS	Non-Intentionally Added Substance
OML	Overall Migration Limit
PhA	Phthalic acid
PA	Polyamide
PET	Polyethylene terephthalate
PVC	Polyvinyl chloride
PIA	Purified isophthalic acid
SSP	Solid-state polymerisation
SML	Specific migration limit
TPA	Terephthalic acid
Tg	Transition temperature
UV	Ultraviolet
U.S.	United States of America

3.1 General Definition

Polyethylene terephthalate (PET) is a thermoplastic obtained from polycondensation of terephthalic acid and ethylene glycol. It belongs to the family of polyesters and can exist as both an amorphous and a semi-crystalline polymer (Webb et al. 2013).

PET was first used in late 1950s for photographic and X-ray films, as well as in flexible packaging. It was later modified and reinforced with glass fibre, and it was stretched in early 1970s using blow-moulding techniques that produced the first oriented three-dimensional structures. This development initiated the rapid exploitation of PET for producing unbreakable bottles (Wyeth et al. 1973–1974).

PET has become the most widely used packaging material worldwide for bottled water and other non-alcoholic refreshment beverages and has gradually replaced polyvinylchloride (PVC) and glass bottles on the market (Andra et al. 2011; Bach et al. 2013; IBWA 2009). In 2015, the global production of PET resins was 27.8

million tonnes which was dominantly used for the manufacture of packaging materials and beverage bottles (Plastics Insight 2016). The reasons for this development are its low cost of production, its chemical and physical stabilities, its transparency, lightweight and good recyclability (Chen et al. 2005). However, in recent years, concerns are rising about the safety of PET food packaging. In fact, the possible migration of chemical compounds from PET bottles into bottled water may pose health risk to consumers. These chemical compounds are called 'non-intentionally added substances' (NIAS), and they are supposed to have potential estrogenic and/or anti-androgenic activities in addition to cancerogenic or toxic features for human beings (Kassouf et al. 2013).

The EU Regulation No 10/2011 on food packaging materials establishes a positive list of the compounds authorised for use in plastic materials intended to come into contact with food, providing migration limits for several molecules (European Commission 2011a). Moreover, the EU Regulation No 1935/2004 gives great importance to NIAS and specifies that '*the notion of risk due to the substance, concerns not only the substance itself but also the impurities in this substance and any reaction or degradation product*' (European Parliament and Council 2004).

3.2 PET Synthesis

PET is particularly suitable for food packaging applications due to its chemical inertness and its physical properties. Usually, it is an amorphous glass-like material, but it can be also in a semi-crystalline form when processed into packaging as a result of crystallinity induced by either heat treatment or the orientation of the polymer chains.

PET is a long-chain polymer that belongs to the family of polyesters (Brody and Marsh 1997). Considering potential chemical combinations involved in its composition, the generic term 'PET' may cover a wide range of polymer structures. It is mainly formed from the monomers terephthalic acid (TPA) or dimethyl terephthalate (DMT) on the one side and monoethylene glycol (MEG) on the other hand. In order to enhance processing and performance, small amounts of co-monomers such as isophthalic acid (IPA) or 1,4-cyclohexanedimethanol (CHDM) may also be incorporated into PET.

The first step in the manufacture of PET is the pre-polymerisation of DMT or TPA with MEG. When heated together, an esterification reaction occurs forming an intermediate compound (monomer) named bis-hydroxyethyl-terephthalate (BHET) and low-molecular-weight polymers (oligomers, with 'n' between 2 and 5, Fig. 3.1). When PET is manufactured by using the pre-polymerisation of TPA, water produced as by-product during this first step is removed (Fig. 3.1). When PET is manufactured by using the pre-polymerisation of DMT, methanol is produced as by-product instead of water. Regardless the molecule used for the pre-polymerisation (TPA or DMT), PET synthesis is due to the polymerisation of the remaining mixture at a temperature of 270–295 °C under vacuum. At this stage, the PET is a viscous, molten liquid that

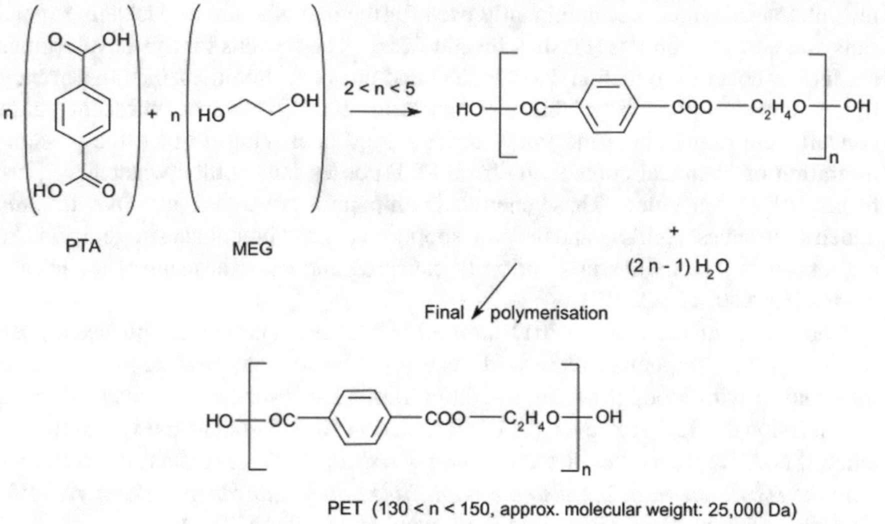

Fig. 3.1 Production of PET from PTA and MEG via initial condensation and oligomerisation (until $n = 2 - 5$), water removal and subsequent PET formation. This picture has been realised by Carmelo Parisi, currently a student at the Liceo Scientifico Stanislao Cannizzaro, Palermo, Italy, with BKchem version 0.13.0, 2009 (http://bkchem.zirael.org/index.html)

is extruded and water quenched into granules to form a solid, glass-like amorphous material.

A second polymerisation step, named solid-state polymerisation (SSP), is conducted at lower temperatures (220–235 °C) than the first polymerisation, either under vacuum or under positive pressure using nitrogen. By-products such as water are eliminated from the interior of pellets to the surface, while volatile impurities or thermal degradation products (e.g. acetaldehyde or residual glycols) are removed by thermal desorption. The SSP process notably reduces the potential quantity of residual substances able to migrate (to food). For this reason, PET is considered a fully compatible food-grade material. Nonetheless, the purity of the starting monomers is the key to achieve the high product quality required for food packaging because once the polymer is formed, it is very difficult to purify except for the removal of volatile materials. Finally, catalysts are used at very low concentrations to promote the reactions. Antimony is the preferred catalyst. Anyway, other catalysts such as titanium, germanium and aluminium may also be used (Welle and Franz 2011).

The conventional process for PET manufacturing is highly energy consuming as well as investment and maintenance intensive. Furthermore, it creates products sensitive to degradation and can easily develop agglomerates with an irregular molecular structure as well as high crystallinity. For this reason, recent technologies have been developed to produce high-molecular-weight PET for food packaging without the SSP step and, consequently, with a conspicuous energy saving. Among these technologies, 'Melt to Resin' (MTR) technology is a continuous polycondensation process for producing PET resin from the feedstock PTA (purified terephthalic acid), PIA (purified isophthalic acid) and EG (ethylene glycol), including conventional co-monomers and additives, in a melt-phase polymerisation process. The technology produces a resin with a final high intrinsic viscosity and an unusually low acetaldehyde content.

3.3 The Manufacture of PET Bottles

The production of PET bottles requires injection moulding of preforms and subsequent stretching and blowing of intermediates into bottles. Injection moulding involves two distinct processes. The first comprises the elementary steps of solids transport, melt generation, mixing, and pressurisation and flow, which are carried out in the injection unit of the moulding machine; the second is the product shaping and 'structuring', which takes place in the mould cavity. The polymer melt is produced in an extruder and exits into a reservoir connected to a hydraulic piston device, which is cyclically pressurised to deliver the melt into the cold mould cavity.

Schematically, the injection blow-moulding process for manufacturing PET bottles is divided into three different steps: injection, stretching and blowing. Amorphous preforms are obtained by processing PET granules. During the injection step, molten polymer flows into the injection cavity via the hot runner block, to produce the desired shape of the preform with a mandrel producing the inner diameter and the injection cavity the outer. Once conditioned to the correct temperature the preform is ready for stretching and blowing to achieve bi-axially oriented bottles.

Good barrier properties and low gas permeability are directly affected by the degree of crystallinity and the orientation of polymer chains in PET bottles. The selection of an adequate blow temperature around 20 °C above the PET glass transition temperature (T_g) is essential to achieve these properties (Tadmor and Gogos 2006). One recent development is the blending of immiscible lamellar polyamide (PA) phases within the PET. This reduces the permeability of O_2 and CO_2 by a factor of two or more. A hexanedioic acid polymer with 1,3-benzenedimethanamine (called MXD-6) is used as barrier material in PET bottles (FSA 2007).

Finally, glass-like transparency is a desired quality for drinking water bottles. To this purpose, co-polymerisation with diethylene glycol (DEG) and isophthalic acid (IPA) is usually carried out to minimise polymer thermal crystallisation during the production of preforms and the blow-moulding process. Both co-monomers reduce the size of spherulites and, as a result, the final container is transparent (Holland and Hay 2002).

3.3.1 PET Additives

Additives are essential components of plastic formulations. They provide mainte-nance of polymer properties, performance and long-term use. They can be classified into two main groups: additives to maintain polymer properties and additives to extend polymer properties.

The first class comprises antioxidants, plasticisers, heat stabilisers, processing aids and lubricants. All these additives are used to transform plastics into the desired shape and to protect the materials from degradation by heat, oxidation or mechanical and chemical attacks. Additives to extend polymer properties help polymer molecular weight remain unchanged for an extended period or under specific use conditions, such as outdoor and ultraviolet (UV) light. Examples of these additives are UV/light stabilisers, antimicrobials, flame retardants and also antioxidants. In addition, some additives are incorporated into polymers with the aim of protecting goods. Examples are UV absorbers and oxygen scavengers to protect packaged food (Pfaendner 2013).

Different additives can be used in different types of polymeric packaging materials.

Plasticisers are used to improve flexibility, workability and stretchability of poly-mers. Some plasticisers used in PET formulations include phthalic esters such as phthalic acid (PhA), bis(2-ethylhexyl) phthalate (DEHP), dimethyl phthalate (DMP), diethyl phthalate (DEP), diisobutyl phthalate (DiisoBP) and dibutyl phthalate (DBP) (Montuori et al. 2008).

Antioxidants are used in concentrations between 0.05 and 1% depending on poly-mer structures and requirements. They are added to slow the onset of oxidative degradation of PET from exposure to UV light and oxygen. In fact, the combined action of light and oxygen results in photo-oxidation with polymeric degradation and irreversible changes of the polymer structure. Butylated hydroxytoluene (BHT), methane (named Irganox 1010) and bisphenolics such as Cyanox 2246 and 425 are the most common phenolics used as antioxidants (Higuchi et al. 2004; Yilmaz and Temel 2016).

Heat stabilisers are added to PET to prevent thermal degradation and to minimise the amounts of acetaldehyde and formaldehyde generated during the processing of bottle preforms (Villain et al. 1995). There are three major types of primary heat stabilisers: mixed metal salt blends, organotin compounds, lead compounds and three secondary heat stabilisers: alkyl organophosphites, epoxy compounds and β-diketones (Kattas et al. 2000). Other heat stabilisers, although more effective if compared with above-mentioned compounds, are not suitable for use in food contact plastics due to their potential toxicity.

Finally, slip compounds are added to reduce the coefficient of friction of poly-meric surfaces. Besides providing lubrication to the film surface, slip agents also impart antistatic and anti-sticking properties and reduce melt viscosity (Sablani and Rahman 2007). Common slip compounds are fatty acid amides (mainly erucamide and oleamide), fatty acid esters, metallic stearates (for example, zinc stearate) and waxes (Bhunia et al. 2013).

3.4   Food Packaging Legislation in the EU

In Europe, the Directive 98/83/EC is related to the quality of water intended for human consumption, while the EU Regulation No 1935/2004 and the EU Regulation No 10/2011 set out general principles of safety and inertness for all food contact materials (FCM) and rules on the composition of plastic FCM, respectively.

FCM are widely used in everyday life in the form of food packaging, kitchen utensils, etc. They are made of substances that might transfer their chemical constituents into the food. The migration of FCM substances into food could bring changes in food safety and in food quality. Moreover, if ingested in large quantities, FCM chemicals might endanger human health. The Regulation (EC) No 1935/2004 aims at ensuring FCM safety but also the effective functioning of the internal market in FCM goods.

Currently, only four FCM out of the 17 FCM listed in Annex I of the Regulation (EC) No 1935/2004 for which specific safety requirements may be adopted are subject to harmonisation at the EU level, namely:

(a)   Plastics (including recycled plastics),
(b)   Ceramics,
(c)   Regenerated cellulose film and
(d)   Active and intelligent materials.

The principles set out in Regulation (EC) No 1935/2004 require that materials do not

- Release their constituents into food at levels harmful to human health.
- Change food composition, taste and odour in an unacceptable way.

Moreover, the framework provides

- Special rules for active and intelligent materials.[1]
- Additional EU measures for specific materials (e.g. for plastics).
- The procedure to perform safety assessments of substances used to manufacture FCM
- Rules on labelling including an indication for use.
- Documentation and traceability.

Specific measures may include lists of substances authorised for use in the manufacturing of materials and articles; purity standards for substances put on the positive lists; special conditions of use for substances on the positive lists and/or the materials and articles in which they are used; specific and/or overall limits on the migration of certain constituents into the food; basic rules for checking compliance with the harmonised rules; etc.

---

[1] *Materials and articles, including active and intelligent materials and articles, shall be manufactured in compliance with good manufacturing practice so that, under normal or foreseeable conditions of use, they do not transfer their constituents to food in quantities which could endanger human health, or bring about an unacceptable change in the composition of the food, or bring about a deterioration in the organoleptic characteristics thereof* (European Parliament and Council 2004, Article 3).

Furthermore, the Regulation (EC) No 1935/2004 lays down rules as regards the labelling of food contact materials when placed on the market. The labelling requirements play an important role in terms of traceability of food contact materials and articles. In particular, each FCM should be accompanied by the words 'for food contact', or a specific indication of their use (e.g. wine bottle), or the so-called 'wine glass and fork' symbol reproduced in Annex II. If necessary, special instructions to be observed for safe and appropriate use shall be indicated.

Finally, to ensure traceability of the material or article, the name and registered office of the manufacturer, processor or seller responsible for placing on the internal market shall be indicated. The required information should be clearly legible and inedible.

Commission Regulation (EC) No 2023/2006 defines Good Manufacturing Practice (GMP). GMP applies to all food contact materials and all stages of manufacture, processing and distribution of materials and articles, excluding the production of starting substances that is covered by other legislations. Under the GMP Regulation, quality assurance and quality control systems are mandatory for interested business operators. Also, printing inks specifically used on to the non-food contact side of FCM are ruled.

Regulation (EC) No 450/2009 establishes a Union list of substances permitted for the manufacture of active and intelligent materials and articles. 'Active materials and articles' are intended to extend the shelf life or to maintain or improve the condition of packaged food. They are specifically designed to incorporate components that would release or absorb substances into or from the packaged food or the environment surrounding the food (e.g. moisture absorbers or oxygen scavengers). 'Intelligent materials and articles' are materials and articles which monitor the condition of packaged food or the environment surrounding the food (e.g. time temperature indicators). Regulation (EC) No 450/2009 lays down specific rules for the so-called 'functional barrier', which means barrier consisting of one or more FCM layers that ensure that the finished material or article complies with Article 3 of Regulation (EC) No 1935/2004. This barrier is a layer within the FCM or articles preventing the migration of substances from behind that barrier into the food. The maximum tolerated migration level is 0.01 mg of substance/kg food.

Polymers exerting the function of 'active oxygen barrier' are used for PET-bottled waters in order to prevent the permeation of oxygen through the PET article. The substance can be incorporated into the primary packaging (e.g. into bottle wall). In the case that the internal oxygen is absorbed too, but such an effect is unintentional and rather minimal, it is not considered as an active packaging under Regulation (EC) No 450/2009. If there is an intentional effect, the application is covered by the definition and it should be declared and proven in the application.

Finally, absorbers able to exert their effect on substances that constitute the packaging material and not on the food or the environment surrounding the food are not covered by the definition. Such absorbers may have the aim to suppress the release of those substances from the packaging into the food or the environment surrounding the food. An example of the above may be an absorber of acetaldehyde from PET.

There are also specific rules on some starting substances used to produce FCM. The most comprehensive specific EU measure is the Commission Regulation (EU) No 10/2011 on plastic materials and articles (European Commission 2011a, b). The Regulation sets out rules on the composition of plastic FCM and establishes a Union List of substances that are allowed for use in the manufacture of plastic FCM. The Regulation also specifies restrictions on the use of these substances and sets out rules to determine the compliance of plastic materials and articles, with the definition of 'specific migration limits' (SML) concerning the maximum amount of substances which could be found into the contained food because of migration phenomena. For the substances on the Union list, the SML are established by the EFSA on the basis of toxicity data of each specific substance. To ensure the overall quality of the plastic, the overall migration to a food of all substances together may not exceed the overall migration limit (OML) of 60 mg/kg food, or 10 mg/dm^2 of the contact material.

To ensure the safety, quality and compliance of plastic materials (including recycled plastics), active and intelligent materials, ceramics and regenerated cellulose film, adequate data on the composition of materials has to be communicated via the manufacturing chain, up to but not including the retail stage. For this purpose a 'Declaration of Compliance' (DoC) needs to be provided. At the moment, DoC is not mandatory for certain types of food contact materials, such as, for example, paper or inks. China has similar requirements on DoC for food contact materials and containers.

Other Regulations exist to control the recycling processes or to regulate specific substances (Table 3.1).

Finally, Commission Recommendation (EU) No 2019/794 of 15 May 2019 sets out a coordinated control plan with a view to establishing the prevalence of certain substances migrating from materials and articles intended to come into contact with food (results of controls in 2019 will be reported in early 2020). For example, in addition to the use of phthalates, other non-phthalate substances may be used in FCM as plasticisers.

Epoxidised soybean oil as well as 1,2-cyclohexane dicarboxylic acid diisononyl ester and terephthalic acid, bis(2-ethylhexyl)ester are authorised for use in the manufacture of plastic FCM and are included in the group SML of 60 mg/kg. As there is little or no available information on their migration into food, it is appropriate to check for the prevalence of these substances migrating into food (European Commission 2011b).

3.5   Food Packaging Legislation in the U.S.

The Food and Drug Administrations (FDA) is the governmental agency in the U.S. that regulates food contact substances, materials and articles. All existing regulations are published in the Code of Federal Regulation (CFR) and all food additives regulations are included in Title 21 CFR parts 170-199.

The U.S. FDA considers three different types of food additives:

Table 3.1 EU Regulations related to recycled FPM. The ambit of application may concern recycling processes and/or regulate specific substances and FPM

European Legislation	Ambit of application
Commission Regulation (EC) No 282/2008 of 27 March 2008 on recycled plastic materials and articles intended to come into contact with foods and amending Regulation (EC) No 2023/2006	General FPM
Council Directive 84/500/EEC of 15 October 1984 on the approximation of the laws of the Member States relating to ceramic articles intended to come into contact with foodstuffs	Ceramic FPM
Commission Directive 2007/42/EC of 29 June 2007 relating to materials and articles made of regenerated cellulose film intended to come into contact with foodstuffs	Regenerated cellulose FPM
Commission Regulation (EU) 2018/213 of 12 February 2018 on the use of bisphenol A in varnishes and coatings intended to come into contact with food and amending Regulation (EU) No 10/2011 as regards the use of that substance in plastic food contact materials	Bisphenol A in FCM
Commission Regulation (EC) No 1895/2005 of 18 November 2005 on the restriction of use of certain epoxy derivatives in materials and articles intended to come into contact with food	Epoxy derivatives in FPM
Commission Directive 93/11/EEC of 15 March 1993 concerning the release of the N-nitrosamines and N-nitrosatable substances from elastomer or rubber teats and soothers	N-nitrosamines and N-nitrosatable compounds in FPM containing elastomer or rubber teats and soothers

(a) Direct food additives—components added directly to the food for a functional purpose (e.g. food colouring), not regulated by food contact regulations.

(b) Indirect food additives—substances that may come into contact with food as part of packaging or processing equipment, but are not intended to be added directly to food and do not have a functional purpose: food contact substances (FCS).

(c) Secondary direct food additives—components that are added to the food due to food treatment (e.g. ionic resins, solvent extraction, etc.).

In particular, 21 CFR 170-190 is focused on testing of food contact materials to check for compliance with the regulations. In most cases, it is mainly focused on the starting materials. The following parts of 21 CFR pertain to 'indirect' food additives:

• 21 CFR Part 175—Adhesives and components of coatings.
• 21 CFR Part 176—Paper and paperboard.
• 21 CFR Part 177—Polymers.

- 21 CFR Part 178—Adjuvants, production aids and sanitisers.

Part 177 is titled 'Indirect food additives: polymers'. Food contact materials used in the U.S. must comply with 21 CFR 177 regulations. The 21 CFR 177.1630 part refers to polyethylene terephthalate materials and their copolymers, which are used in a wide variety of food contact materials from storage containers to flexible packaging. The FDA requires that PET resins intended for use in food contact applications must meet certain requirements for identity and extractable content.

The testing consist in most cases of migration of the granulate material or sheet form of the base polymer or extraction tests. Viscosity of the base polymer may also be required. Sometimes testing is required for the residual content of a monomer (e.g. styrene). Anyway, each polymer type has its own set of testing and limitations that must be met.

FDA regulates also the packaging and labelling of food in order to enhance the safety of food distributed throughout the United States and keep consumers informed about the food. Packaging materials like plastics, coatings, papers, food colourants and adhesives must be regulated and declared safe for use. To this purpose, the FDA classifies any packaging that comes into direct contact with food as a food contact substance (FCS). It is the responsibility of the Office of Food Additive Safety and The Center for Food Safety and Applied Nutrition (CFSAN) to ensure the safety of these food contact substances.

The Environmental Protection Agency (EPA) requires that antimicrobial technology be built into plastic and textiles used in food packaging to prevent the growth of bacteria, mould, discolouration and odour. In addition, the FDA also requires that package labelling includes nutrition guidelines, expiration or '*best if used by*' dates, handling and preparation instructions, and the packaging company's contact information. Furthermore, all food packaging authorisations require an 'Environmental Assessment' (EA): this document is publicly available and it aims at the evaluation of food and feed additives potentially able to damage the environment. An EA serves to provide sufficient evidence and analysis for an agency to determine whether significant environmental impacts may occur from the action. If FDA's Center for Veterinary Medicine determines that the information in the EA demonstrates that no significant environmental impacts are expected, then a 'Finding of No Significant Impact' (FONSI) is prepared. A FONSI is a document that summarises reasons why the agency has concluded that no significant environmental impacts are expected to occur. Otherwise, should unacceptable risks are still expected to occur, risk mitigation options would necessarily be implemented; otherwise, an 'Environmental Impact Statement' (EIS) describing alternative actions will be required.

The FDA mainly reviews the EIS from the packaging material's use and disposal. In particular, they require data on the material's recycling impact. Finally, the FDA's '*Points to Consider for the Use of Recycled Plastics in Food Packaging: Chemistry Considerations*' guidelines state that the use of recycled polymers is permissible if it is of a type previously permitted for food contact, has been kept free of contaminates during the recycling process and the recycled material has been tested to establish suitable purity for reuse in food packaging (Bradley and Coulier 2007; FDA 1992).

References

Andra SS, Makris KC, Shine JP (2011) Frequency of use controls chemical leaching from drinking water containers subject to disinfection. Water Res 45(20):6677–6687. https://doi.org/10.1016/j. watres.2011.10.001

Bach C, Dauchy W, Severin I, Munoz J-F, Etiennne S, Chagnon M-C (2013) Effect of temperature on the release of intentionally and non-intentionally added substances from polyethylene terephthalate (PET) bottles into water: chemical analysis and potential toxicity. Food Chem 139(1–4):672–680. https://doi.org/10.1016/j.foodchem.2013.01.046

Bhunia K, Sablani SS, Tang J, Rasco B (2013) Migration of chemical compounds from packaging polymers during microwave, conventional heat treatment, and storage. Compr Rev Food Sci Food Saf 12(5):523–545. https://doi.org/10.1111/1541-4337.12028

Bradley E, Coulier L (2007) Report FD 07/01: an investigation into the reaction and breakdown products from starting substances used to produce food contact plastics. Food Standards Agency, London

Brody AL, Marsh KS (eds) (1997) The Wiley encyclopedia of packaging tecnology. Wiley, New York

Chen DQ, Wang YZ, Hu XP, Wang DY, Qu MH, Yang B (2005) Flame-retardant and anti-dripping effects of a novel char-forming flame retardant for the treatment of poly (ethylene terephthalate) fabrics. Polym Degrad Stabil 88(2):349–356. https://doi.org/10.1016/j.polymdegradstab.2004. 11.010

Commission European (2006) Commission Regulation (EC) No 2023/2006 of 22 December 2006 on good manufacturing practice for materials and articles intended to come into contact with food. Off J Eur Union L384:75–78

Commission European (2009) Commission Regulation (EC) No 450/2009 of 29 May 2009 on active and intelligent materials and articles intended to come into contact with food. Off J Eur Union L135:3–11

Commission European (2011a) Commission Regulation (EU) No 10/2011 of 14 January 2011 on plastic materials and articles intended to come into contact with food. Off J Eur Union L12:1–89

Commission European (2011b) Commission Recommendation (EU) 2019/794 of 15 May 2019 on a coordinated control plan with a view to establishing the prevalence of certain substances migrating from materials and articles intended to come into contact with food. Off J Eur Union L48:28–32

European Parliament and Council (2004) Regulation (EC) No 1935/2004 of the European Parliament and of the Council of 27 October 2004 on materials and articles intended to come into contact with food and repealing Directives 80/590/EEC and 89/109/EEC. Off J Eur Union L338:4–17

FDA (1992) Points to consider for the use of recycled plastics in food packaging: chemistry considerations. U. S. Food and Drug Administration, Center for Food Safety and Applied Nutrition, Washington, DC

Higuchi A, Yoon BO, Kaneko T, Hara M, Maekawa M, Nohmi T (2004) Separation of endocrine disruptors from aqueous solutions by pervaporation: dioctylphthalate and butylated hydroxytoluene in mineral water. J Appl Polymer Sci 94:1737–1742

Holland BJ, Hay JN (2002) Analysis of comonomer content and cyclic oligomers of poly(ethylene terephthalate). Polymer 43(6):1797–1804

IBWA (2009) Bottled water code practice. International Bottled Water Association (IBWA), Alexandria. https://www.bottledwater.org/files/IBWA%20Code%20of%20Practice% 20Updated%202009%20Final_3.pdf. Accessed 21 Nov 2019

Kassouf A, Maalouly J, Chebib H, Rutledge DN, Ducruet V (2013) Chemometric tools to highlight non-intentionally added substances (NIAS) in polyethyleneterephthalate (PET). Talanta 115:928–937. https://doi.org/10.1016/j.talanta.2013.06.029

Kattas L, Gastrock F, Levin I, Cacciatore A (2000) Plastics additives. In: Modern plastics handbook, 4th edn. McGraw-Hill, New York

Montuori P, Jover E, Morgantini M, Bayona JM, Triassi M (2008) Assessing human exposure to phthalic acid and phthalate esters from mineral water stored in polyethylene terephthalate and glass bottles. Food Addit Contam Part A Chem Anal Control Expo Risk Assess 25(4):511–518. https://doi.org/10.1080/02652030701551800

Pfaendner R (2013) Polymer additives. In: Saldìvar-Guerra E, Vivaldo-Lima E (eds) Handbook of polymer synthesis, characterization, and processing, 1st edn. Wiley, Hoboken. https://doi.org/10.1002/9781118480793.ch11

Plastics Insight (2016) Plastic insight global PET resin production capacity. Plastic Inside, Noida. https://www.plasticsinsight.com/global-pet-resin-production-capacity/. Accessed 21 Nov 2019

Sablani SS, Rahman MS (2007) Food packaging interaction. In: Rahmna MS (ed) Handbook of food preservation. CRC Press, Boca Raton

Tadmor Z, Gogos CG (2006) Principles of polymer processing. Wiley Interscience, 2nd edn. Wiley, Hoboken

Villain F, Coudane J, Vert M (1995) Thermal degradation of polyethlylene terephthalate: study of polymer stabilization. Polym Degrad Stab 49(3):393–397. https://doi.org/10.1016/0141-3910(95)00121-2

Webb H, Arnott J, Crawford R, Ivanova E (2013) Plastic degradation and its environmental implications with special reference to poly(ethylene terephthalate. Polymers 5(1):1–18. https://doi.org/10.3390/polym5010001

Welle F, Franz R (2011) Migration of antimony from PET bottles to beverages: determination of the activation energy of diffusion and migration modeling compared with literature data. Food Addit Contam Part A Chem Anal Control Expo Risk Assess 28(1):115–126. https://doi.org/10.1080/19440049.2010.530296

Yilmaz MA, Temel H (2016) Screening of Polymer Additives in Drinking Water Stored in PET Bottles by UHPLC-ESI-IT-TOF MS. J Environ Sci Eng A 5:59–64. https://doi.org/10.17265/2162-5298/2016.02.001

Chapter 4
Non-Intentionally Added Substances

Abstract Packaged food can contain 'non-intentionally added substances' as a result of the interaction between different ingredients in the packaging materials, from degradation processes and from impurities present in raw materials used for their production. One of the main sources of polymer degradation includes exposure of the polymer to high temperatures and to the ultraviolet–visible light. In the interaction of polyethylene terephthalate bottles with mineral water, several non-intentionally added substances have been identified, such as phthalates, aldehydes (acetaldehyde and formaldehyde) and volatile organic compounds (benzene, toluene, xylene, ethylbenzene, etc.). It is likely that in the majority of cases, due to their very low levels these substances will not be of any health concern.

Keywords Bisphenol A · Endocrine-disrupting chemical · Food contact material · Intentionally added substance · Non-intentionally added substance · Tolerable daily intake · Volatile organic compound

Abbreviations

NP	4-Nonylphenol
ATSDR	Agency for Toxic Substances and Disease Registry
APEO	Alkylphenol Ethoxylates
DEHP	Bis(2-ethylhexyl) Phthalate
BPA	Bisphenol A
BBP	Butyl Benzyl Phthalate
BHT	Butylated Hydroxytoluene
CDC	Centers for Disease Control and Prevention
DNA	Deoxyribonucleic Acid
DBP	Dibutyl Phthalate
DEP	Diethyl Phthalate
DEG	Diethylene Glycol
DMP	Dimethyl Phthalate
EDC	Endocrine-Disrupting Chemicals

EPA   Environmental Protection Agency
EFSA   European Food Safety Authority
EU   European Union
FCM   Food Contact Material
HAA   Hormonally Active Agents
IAS   Intentionally Added Substance
IARC   International Agency for Research on Cancer
MEHP   Mono-2-Ethylhexyl Phthalate
MBP   Mono-n-Butyl Phthalate
NIAS   Non-Intentionally Added Substance
$NPEO_2$ Nonylphenol Diethoxylate
NPEO   Nonylphenol Etoxylated
$NPEO_1$ Nonylphenol Monoethoxylate
NOEL   No-Observed-Adverse-Effect Level
OP   Octylphenol
OML   Overall Migration Limit
PhA   Phthalic Acid
PET   Polyethylene Terephthalate
PVC   Polyvinyl Chloride
SML   Specific Migration Limit
TDI   Tolerable Daily Intake
NPP   Tris(nonylphenyl)phosphite
U.S.   United States of America
VOC   Volatile Organic Compound
WHO   World Health Organization

4.1 Non-Intentionally Added Substances (NIAS)

Plastic packaging contains many substances, such as additives and processing aids, which may migrate into the packaged food. These chemicals are called 'non-intentionally added substances' (NIAS), and may represent impurities of raw materials employed for the production of the packaging material, or may derive from degradation of the material components (Nerin et al. 2013). In the majority of cases, these substances are not dangerous for human health. Nonetheless, in recent years, concerns about the safety of packaged foods have increased noticeably, mainly because migration studies of NIAS released by PET rarely combine chemical analysis with toxicological assessments. Therefore when bioassays demonstrate positive responses, analytical data to identify the responsible compounds are always lacking and conclusions are difficult to draw (Brüschweiler et al. 2014; Hollnagel et al. 2014).

The European Regulation (EC) No 1935/2004 sets up the criteria which aim to provide safely use of food contact materials (FCM) and articles in the European Union (EU). The Regulation establishes that individual regulations for different groups of materials (e.g. plastics, paper, metals and alloys, adhesives, printing inks, etc.) can be adopted at the EU level. Plastic materials and articles intended to come into contact with food are regulated by Commission Regulation (EU) No 10/2011. This Regulation recognises that NIAS can be formed during the manufacture and use of plastic materials. Therefore, only substances that are authorised may be intentionally used in the manufacture of plastic materials and articles. These substances (authorised monomers, other starting substances, macromolecules obtained from microbial fermentation, additives and polymer production aids) are listed in the 'Union list', which is a positive list of 885 compounds authorised for use in plastic formulations and manufacturing. These listed substances, called 'intentionally added substances' (IAS), can be used to manufacture plastic materials, with the restrictions and specifications established in the list. In particular, the EU Regulation No 10/2011 applies to materials and articles and parts thereof consisting exclusively of plastics, plastic multi-layer materials and articles held together by adhesives, plastic layers or plastic coatings forming gaskets in caps and closures, as well as plastic layers in multi-material multi-layer materials and articles.

Since it is impossible to list and consider all impurities in the authorisation, the EU Regulation No 10/2011 gives the responsibility to the manufacturers for risk analysis. Therefore, substances not included in the Union list may be present in a material or article.

Substances used in the manufacture of FCM are regulated with maximum limits that may migrate into foodstuffs without causing any health concerns. The EU Regulation No 10/2011 provides two migration limits for quite a number of molecules: specific migration limits (SML) for individual authorised substances fixed on the basis of a toxicological evaluation, and the overall migration limit (OML) for all substances that can migrate from FCM to foods.

NIAS formation is frequently due to degradation processes following the exposure of the polymer to high temperatures, ultraviolet and visible light exposure, humidity, atmospheric components and contact with liquids. All these events may be responsible for chemical and physical ageing of PET bottles provoking the formation of scratches and cracks on its surface. As a direct consequence, formation of micro-cavities occurs. During these processes, new molecules with a lower molecular weight can be formed, and these molecules will have higher diffusion coefficients and therefore a higher migration potential.

Degradation can take place in the polymer itself or in the additives used for improving its physicochemical characteristics. NIAS coming from polymer degradation processes are generally related to carbonyl compounds, mainly formaldehyde and acetaldehyde, coming from the thermal degradation of polyethylene terephthalate (PET) (Mutsuga et al. 2006). Ethylene terephthalate dimers and trimers were also determined as degradation compounds from recycled PET (Bentayeb et al. 2007). Some additives, such as antioxidants or light stabilisers, added to the polymer can also be degraded. In addition, the degradation process of additives added to adhesives, or coatings, or inks used in food packaging manufacture can provide NIAS to the packaging material and their subsequent migration into the food (Canellas et al. 2010). Another frequent reason for finding NIAS in migration from food packaging is the presence of impurities coming from the raw materials used during the polymer manufacturing.

4.2 NIAS and PET

Polymers constitute the main structural component of plastic materials. They are produced by a process known as 'polymerisation' (Selke 2005). The starting substances of polymerisation are mainly monomers, which react with other substances to make polymers. The intermediate mass can be moulded under pressure and heating, and turned into different containers and other food-related applications (Krochta 2007). Polymer additives, such as antioxidant, solvents, plasticisers, thermal and light stabilisers are also used to improve plastic flexibility and polymer resistance to degradation by heat and light (Arvanitoyannis and Kotsanopoulos 2014; Krochta 2007). Acrylic adhesives used to stick labels on the packages or to form the geometric shape of the package are also a possible source of NIAS which can migrate from PET to contained water (Canellas et al. 2010). In addition, organic or inorganic materials can be used as printing agents to coat plastic materials and articles.

Each step of the manufacture of PET bottles can introduce and generate NIAS in the polymer, posing a risk of unacceptable migration of chemical substances from PET bottles into water in contact. External contamination from the surrounding environment and the influence of storage conditions should be considered also. High temperatures and the presence of oxygen in PET can promote thermo-mechanical and thermo-oxidative reactions generating numerous NIAS in the polymer and contributing to change its chemical structure. In particular, chemical compounds such as aldehydes (acetaldehyde, formaldehyde and benzaldehyde), aliphatic hydrocarbons (C1–C4), aromatic hydrocarbons (benzene, toluene, ethylbenzene, xylene and styrene) and esters (vinyl benzoate, methyl acetate) have been recovered in PET bottles subjected to generally applied temperatures in PET production and processing (between 200 and 300 °C). In the interaction of PET bottles with mineral water, several substances have been identified, such as phthalates, aldehydes (acetaldehyde and formaldehyde) and volatile organic compounds (benzene, toluene, xylene, ethylbenzene, etc.).

4.2.1 Phthalates

Phthalates, diesters of 1,2-benzenedicarboxylic acid (phthalic acid), are a large group of chemicals employed in several products, such as toys, vinyl flooring and wall covering, detergents, lubricating oils, food packaging, inks, pharmaceuticals, blood bags and tubing, as well as personal care products, such as nail polish, hair sprays, aftershave lotions, perfumes, soaps and shampoos (Serrano et al. 2014).

Nowadays, they are not thought to be used in the manufacture of PET bottles (ILSI 2000). Despite this, the analysis of PET reveals phthalates in PET-bottle materials, which can migrate into bottled drinking water. The origin of these compounds has not been clearly established, and it has been argued that they could come from cap-sealing resins, background contamination, water processing steps or recycling processes) (Grob et al. 2006; Muncke 2009). Several studies have demonstrated that the possibility of the presence of phthalates in PET-bottled water is due to factors including storage duration, temperature change and sunlight. Anyway, although there is an increasing demand for more comprehensive studies on the role of possible factors in the migration of phthalates into PET-bottled water under common usage conditions, it seems that storage at low temperatures (refrigerator and freezing conditions) especially when compared to high temperatures (>25 °C) does not cause significant migration of these chemicals. Low storage temperatures and short storage times are preferable for mitigating potential release of phthalates from plastic bottle to water, thus reducing human daily intakes of phthalates through water consumption (Greifenstein et al. 2013).

Dibutyl phthalate (DBP) is used primarily as a specialty plasticiser for nitrocellulose, polyvinyl acetate and polyvinyl chloride. It has been used in plastisol formulations for carpet back coating and other vinyl compounds. DBP is widespread in the environment and has been identified at low levels in air, water and soil. Therefore, humans may be exposed to DBP by inhalation of air or by ingestion of water or food containing DBP. Although adverse effects on humans from exposure to DBP have not been reported, eating large amounts of DBP can affect animals' ability to reproduce (ATSDR 1990).

Di-2-ethylhexyl phthalate (DEHP) is principally used as a plasticiser in the production of polyvinyl chloride (PVC) and vinyl chloride resins. PVC is used in many common items such as toys, adhesives, coatings, paper and paperboard. PVC is also used to produce disposable medical examination and surgical gloves, the flexible tubing used to administer parenteral solutions, and the tubing used in hemodialysis treatment. Non-plasticising uses include the use of DEHP as a solvent in erasable ink.

DEHP is a ubiquitous environmental contaminant. The principal route of human exposure to DEHP is ingestion of contaminated food, especially fish, seafood or fatty foods, with an estimated daily dose of about 0.25 mg. The highest exposures to DEHP result from medical procedures such as blood transfusions or hemodialysis, during which DEHP may leach from plastic equipment into biological fluids. Workers in industries manufacturing or using DEHP as plasticiser may be frequently exposed to above average levels of this compound. Eating high doses of DEHP for a long time resulted in liver cancer in rats and mice. The U.S. Department of Health and Human Services has determined that DEHP may reasonably be anticipated to be a carcinogen (ATSDR 1993a). The International Agency for Research on Cancer (IARC) designated DEHP to 'Group 2B' (possibly carcinogenic to humans) (IARC 1982). Short-term exposures to DEHP interfered with sperm formation in mice and rats. After long-term exposure, fertility of both male and female rats was decreased.

Studies of pregnant mice and rats exposed to DEHP resulted in effects on the development of the foetus, including malformation of foetus and reduction in neonatal weight and survival. Long-term exposure of animals to DEHP resulted in structural and functional changes in the kidneys (ATSDR 1993a, b).

Dimethyl phthalate (DMP) shows low toxicity, but when accidentally ingested in large amounts it may cause gastrointestinal irritation, central nervous system depression with coma, and hypotension. It is also an irritant to the eyes and the mucous membranes (Clayton and Clayton 1981).

Diethyl phthalate (DEP) is needed as a plasticising agent in many industrial applications. Human exposure to DEP can result from breathing contaminated air, eating foods into which DEP has leached from packaging materials, eating contaminated seafood, drinking contaminated water or as a result of medical treatment involving the use of PVC tubing (e.g. dialysis patients). Adverse effects on humans from exposure to DEP have not been reported. On the other hand, DEP can cause death in animals and in newborn rats provided that ingestion is remarkable (by mouth) or mothers assume approximately 3 g/kg of DEP during pregnancy, while brief exposure and low ingestion levels do not apparently cause harmful damages (ATSDR 1993b).

DBP, butyl benzyl phthalate (BBP), and DEHP are classified as 'endocrine-disrupting chemicals' (EDC) or 'hormonally active agents' (HAA) due to their anti-androgenic or pro-estrogenic effects. An endocrine disruptor is defined as '*an exogenous chemical, or mixture of chemicals, that can interfere with any aspect of hormone action*' (Zoeller et al. 2012). These chemicals can bind to the endocrine receptors to activate, block or alter natural hormone synthesis and degradation which results in false lack or abnormal hormonal signals that can increase or inhibit normal endocrine function (Zoeller et al. 2012). DiNP, BBP and DEHP are also weakly estrogenic. Some metabolites of phthalates such as mono-2-ethyl-hexyl phthalate (MEHP), mono-*n*-butyl phthalate (MBP) and monoethyl phthalate are also capable of disturbing the hormonal activity European Commission 2001).

Furthermore, DEHP, DBP and DiNP are suspected carcinogens, as well as toxic to liver, kidneys (Gomez-Hens and Aguilar-Caballos 2003) and reproductive organs (Swan et al. 2005). Tolerable daily intake (TDI) values established by the European Food Safety Authority (EFSA) panel for BBP, DBP and DEHP are 500, 10 and 50 µg/kg/body weight/day, respectively (EFSA 2019). A TDI has not yet been defined for DiBP.

4.2.2   Aldehydes

Aldehydes, obtained from alcohols by means of hydrogen removal (dehydrogenation), undergo several chemical reactions, including polymerisation. Their combination with other types of molecules produces the so-called aldehyde condensation polymers, which are used in plastics.

Formaldehyde and acetaldehyde are the dominant carbonyl compounds identified in water (Dabrowska et al. 2004; Nawrocki et al. 2002). They are highly volatile and can migrate from bottles into water after filling and storage and, consequently, this could lead to a change in taste and odour of the bottled drinking water (Dabrowska et al. 2004; Nawrocki et al. 2002). In particular, acetaldehyde may cause an undesirable sweet and fruity taste in bottled drinking waters (Mutsuga et al. 2006).

Sources of aldehydes in bottled water include oxidative water treatment processes such as ozonation and chlorination. Also, the influence of thermo-oxidative and chemical degradation of PET containers has been extensively studied (Romão et al. 2009; Zhang and Ward 1995). In bottled water, migration or formation of aldehydes from the PET containers may be due to a degradation process under environmental factors as sunlight and ultraviolet (UV) light (Bach et al. 2012; Mutsuga et al. 2006; Ozlem 2008). Therefore, temperature as well as storage conditions of PET-bottled waters needs to be controlled.

4.2.2.1 Acetaldehyde

Acetaldehyde is a widespread, naturally occurring, colourless and flammable liquid with a suffocating smell. It is found in various plants, fruits, vegetables, cigarette smoke, gasoline and diesel exhaust. The main use of acetaldehyde is as an intermediate for the synthesis of other chemicals. It is used in the production of perfumes, polyester resins and drugs. Acetaldehyde is also used as a solvent in the rubber, tanning and paper industries, as a fruit and fish preservative, as a flavouring agent, for hardening gelatin, as a denaturant for alcohol and in fuel compositions.

Acute exposure to its vapours results in irritation of eyes, skin and the respiratory tract. Symptoms of chronic intoxication of acetaldehyde resemble those of alcoholism. The International Agency for Research on Cancer (IARC) has classified acetaldehyde as a 'Group 2B' carcinogen to humans, based on inadequate human cancer studies and animal studies that have shown nasal tumours in rats and laryngeal tumours in hamsters. Acetaldehyde associated with consumption of alcoholic beverages and formed from ethanol endogenously is a 'Group 1' carcinogen to humans (IARC). However, also foodstuffs and 'non-alcoholic' beverages containing less than 2.8% ethyl alcohol may expose the oral cavity and oesophagus to acetaldehyde. Moreover, depending on the gastric emptying rate, the mucosa of *Helicobacter pylori*-infected or chlorhydric stomachs may be exposed by the same mechanism to acetaldehyde daily for hours. Therefore, many foodstuffs and 'non-alcoholic' beverages are important but unrecognised sources of local acetaldehyde exposure (Salaspuro 2011).

4.2.2.2   Formaldehyde

Formaldehyde is a colourless gas synthesised by methanol oxidation. It is used as general-purpose chemical reagent for laboratory applications, as well as a disinfectant and antiseptic, although its presence is reported currently in environmental atmosphere as a result of anthropic pollution. In drinking water, formaldehyde arises mainly from the oxidation of natural organic matter during chlorination (Becher et al. 1992) and ozonation (Glaze et al. 1989). It also enters drinking water via leaching from polyacetal plastic fittings in which the protective coating has been broken (Liteplo et al. 2002).

Formaldehyde has shown evidence of mutagenicity in prokaryotic and eukaryotic cells in vitro. It binds readily to proteins, ribonucleic acid (RNA) and single-stranded deoxyribonucleic acid (DNA) to induce DNA–protein cross-links and breaks in single-stranded DNA (Ma and Harris 1988).

In vivo, formaldehyde increases both DNA synthesis and the number of micronuclei and nuclear anomalies in epithelial cells in rats (Migliore et al. 1989; Overman 1985). In humans, irritation and allergic contact dermatitis have been associated with exposure of the skin to high levels of formaldehyde. In addition, there is some evidence that formaldehyde is a carcinogen in humans exposed by inhalation. Several studies have provided evidence that formaldehyde may pose a carcinogenic risk of lung or sino-nasal cancer, and possibly lymphoid leukaemia in occupationally exposed groups (IARC 2006). On the basis of studies in which humans and experimental animals were exposed to formaldehyde by inhalation, IARC has classified formaldehyde in the 'Group 1' (carcinogenic to humans).

In water, formaldehyde is highly soluble, and it is largely in the form of methylene glycol and its oligomers (Liteplo et al. 2002). When ingested formaldehyde is readily absorbed by the gastrointestinal tract but it is not carcinogenic by the oral route. Effects in the tissue of first contact following ingestion seem to be related to the concentration of consumed formaldehyde than to its total intake. At present, ingested formaldehyde is reported to be tolerable until 2.6 mg/l on the basis of the 'no-observed-adverse-effect level' (NOEL) of 260 mg/l for histopathological effects in the oral and gastric mucosa of rats administered formaldehyde in their drinking water for 2 years (Liteplo et al. 2002). The Directive 2002/72/EEC has recommended that migration levels of formaldehyde and acetaldehyde in water stored in PET bottles should not exceed 15 and 6 mg/kg, respectively (Directive 2002/72/EEC).

4.2.3   Volatile Organic Compounds

Volatile organic compounds (VOC) are compounds that contain, at least, the element carbon and one or more element from hydrogen, oxygen, sulphur, phosphorus, silicon, nitrogen or halogens, and have a high vapour pressure at room temperature. They are generally referred to as the highly reactive and/or toxic organics emitted by both human-made and natural sources due to their high volatility at normal atmospheric conditions (Godish et al. 2015).

VOC such as benzene, xylene, toluene, ethylbenzene, o-, m- and p-xylenes, as well as styrene, chlorobenzene and benzaldehyde may be present in both bottled waters (and subsequently in contained waters!) due to their non-intentionally addition to PET during its production (Ikem 2010; Ji et al. 2006; Li et al. 2001). The levels of styrene, toluene and xylenes have been found to increase in PET bottles with storage time, while no effect has been detected due to changes in the storage temperature (Alsharifi et al. 2009). On the contrary, the rate of benzene in PET increases exponentially with the applied temperature (Dzieciol and Trzeszczynski 1998). Thus, the overall VOC migration from walls of PET bottles into water increases with storage time and exposure to high temperatures and UV light. Also, the presence of oxygen and humidity can promote degradation reactions and, consequently, VOC release (Di Felice et al. 2008; Ozlem 2008).

With relation to another degradation residue of PET formatting, benzaldehyde in virgin bottles can be found in spite of its absence in the initial formula (granules). Other VOC identified in virgin PET include butoxybenzene, dimethyl cyclohexane-1,4-dicarboxylate, and toluene. In addition, the detection of benzaldehyde in recycled PET-bottled mineral water is an indicator that the used PET is totally or partly recycled, because benzaldehyde is used as solvent for printing inks used for bottles labelling (Ghanem et al. 2013; Gramshaw et al. 1995).

The manufacture of PET involves three main steps, each having the potential for introducing VOC migratory components. The first stage involves the manufacture of ethylene glycol, terephthalic acid and/or dimethyl terephthalate, all from crude oil using catalysts, pressure and heat. In the case of the latter two compounds, p-xylene from the naphtha fraction of crude oil is either oxidised to terephthalic acid or oxidised and esterified (with methanol) to produce dimethyl terephthalate. Ethylene glycol is manufactured by oxidation of ethylene from the gas fraction of crude oil to ethylene oxide (oxirane), which is subsequently hydrolysed with water. The oxidation of ethene to oxirane takes place in the presence of a silver catalyst. The potential migratory components resulting from this step are VOC such as p-xylene, ethylene glycol and p-toluic acid (Freire et al. 1998). Anyway, levels of these volatiles are generally low, indicating no hazard to public health (Kim et al. 1990; Morelli-Cardoso et al. 1997).

It has to be noted that benzene, trihalomethanes and other VOC are reported to influence human health with carcinogenic effects. Due to their toxicity (carcinogenic or mutagenic properties) even at trace levels, the U.S. EPA (2009) set maximum concentration levels (MCL) for potential VOC in drinking water. These values are relatively low especially for benzene which are of 5 and 10 µg/l for EPA and WHO, respectively.

4.3   Evidence of NIAS and Their Source

According to EU Regulation No 1935/2004, '*food contact materials must not transfer their constituents to food in quantities which could endanger human health*' (European Parliament and Council 2004).

PET production does not need antioxidants and plasticising agents, while colour agents are really reduced if compared with the main plastic material. In addition, acetaldehyde scavengers can reduce the formation of acetaldehyde during the melt-process (EFSA 2012). In addition, starting substances and additives are strictly regulated by EU Regulation No 10/2011. Nonetheless, migration of NIAS from PET bottles to water may occur for several reasons.

PET can be degraded due to several exposure factors under normal conditions of use, such as heat and UV light. In addition, certain physicochemical properties of bottled water, such as inorganic composition, carbonation or bacterial presence, influence the leaching of constituents from PET bottles into water (Bach et al. 2013). The most frequent sources of NIAS in PET-bottled water are: (i) degradation processes, (ii) impurities, (iii) neo-formed compounds and (iv) contaminants.

Degradation can take place in the polymer itself or in the additives used for improving PET physicochemical characteristics. One of the main sources of polymer degradation includes exposure of the polymer to high temperatures and to the ultraviolet–visible light. When exposed to sunlight PET may undergo to chemical and physical ageing that provoke the formation of tiny scratches, small cracks and micro-cavities on the surface. During these processes, new molecules with a lower molecular weight and a higher diffusion coefficient can be formed and migrate into the water (Harvey 2012).

The majority of NIAS coming from thermo-mechanical and thermo-oxidative degradation of the polymer are related to carbonyl compounds, where the most relevant compounds are formaldehyde and acetaldehyde, which are generally responsible for an off-flavour (Bach et al. 2014; Osorio et al. 2018). Recently, two compounds not included in the EU Regulation positive list, 2,4-di-*tert*-butylphenol and bis(2-hydroxyethyl)terephthalate, have been identified in PET-bottled water exposed to high temperature. Anyway, these new NIAS were not found to be cytotoxic or genotoxic, and did not show any estrogenic or anti-androgenic activity (Bach et al. 2013).

Some additives such as antioxidants or light stabilisers added to polymers for improving their properties can also be degraded by oxidising conditions. This fact is not surprising because the purpose of these compounds is to act when oxidising conditions such as high temperatures occur, and their mode of action is often to be oxidised earlier than the material. Anyway, degradation products of the most commonly used antioxidants Irganox 1010 and Irgafos 168 were not detected during conventional heating of polypropylene packaging (Alin and Hakkarainen 2011).

Also, additives added to adhesives, coatings or inks used in bottled water manufacture can undergo degradation processes and provide NIAS. The formation of intermediate compounds such as 4b-8-dimethyl-2-isopropylphenanthrene, dehydroabietin,

1-methyl-10,18-bisnorabieta-8,11,13-triene, retene, dehydroabietal and dehydroabietic acid methyl ester can be due to the thermal degradation of abietic acid used as a tackifier in hot melt adhesives for food packaging (Aznar et al. 2011). The acute oral toxicity of abietic acid and dehydroabietic acid is low. Nonetheless, allergenic effects such as airborne contact dermatitis or asthma and rhinitis-like reactions with or without asthma in workers who had been exposed to abietic acid vapours, and its oxidation products for extended periods of time have been reported (The MAK-Collection 2013).

Another frequent reason for NIAS migration from food packaging is the presence of impurities coming from the raw materials or additives used during polymer manufacturing, or coming from adhesive additives used in food packaging.

Several non-volatile compounds were found as potential migrants from adhesives used in food packaging. In particular, among the nonylphenol etoxylated (NPEO) polymers, nonylphenol diethoxylate ($NPEO_2$) was found. This compound is considered the most persistent and toxic NPEO metabolite together with nonylphenol monoethoxylate ($NPEO_1$) (Canellas et al. 2010).

In PET bottles, NIAS can origin also from polyurethane adhesives, which are formed by the polymerization of polyols and diisocyanate monomers. If the ingredients have been mixed wrongly, the remaining unpolymerised aromatic isocyanates will come into contact with water inside the bottle, producing primary aromatic amines (PAA). PAA are very toxic compounds and are suspected to be carcinogenic (Pezo et al. 2012). To protect consumers' health, the EU legislation has established a specific migration limit of 10 ng for total PAA per gram of food (Commission Directive 2007/19/EC).

Inorganic species may be present as residues from catalysts or additives used to produce PET. Antimony is the most relevant NIAS from PET bottles. Antimony (Sb) leaching from PET into water increases rapidly during the first storage period and is thermally activated. In addition, antimony dissolution rate into water is reported to be lower in still water if compared with sparkling water, probably because of low pH values. In contrast, PET exposure to sunlight appears to be less significant than other factors in Sb migration (Cheng et al. 2010; Keresztes et al. 2009). However, it has been seen that only a small fraction of the antimony contained in PET is released into the water (Welle and Franz 2011). Anyway, the SML prescribed for antimony is low (0.04 mg/kg) (European Commission 2011).

Small amounts of antioxidants can be added to the polymer before it is processed to inhibit or reduce the oxidation of plastic material. In food packaging manufacture, alkylphenols, such as 4-nonylphenol (NP) and octylphenol (OP), can be generated by oxidation of the additive tris(nonylphenyl)phosphite (TNPP) or by degradation of alkylphenol ethoxylates (APEO), which are cleaning agents in bottle manufacturing (Casajuana and Lacorte 2003; McNeal et al. 2000). NP and OP are known to be endocrine disrupters (Loos et al. 2007).

Anyway, NP and OP are usually found at low levels in PET-bottled waters and SML have not been established. Also, butylated hydroxytoluene (BHT), a phenolic antioxidant used as a thermostabiliser, is usually recovered in PET-bottled waters at concentrations much lower than the established SML of 3 mg/kg (Commission regulation (EU) No 10/2011; Sheftel 2000).

Finally, when recycled materials are used for food packaging, the possible migration of NIAS coming from contamination of the packaging shall be considered. These contaminants may be chemical compounds coming from the previously packaged food, but also substances resulting from the misuse of the packaging by the consumer before discarding it (Welle 2011). The presence of intrinsic contaminants from the recycling process such as chemical additives and their degradation products also has to be considered. An example of these compounds is bisphenol A—2,2-bis(4-hydroxyphenyl)propane, also named BPA—which has been well characterised as an endocrine disruptor (Elobeid and Allison 2008; Vom-Saal and Myers 2008), potentially leading to reproductive defects, cancer, obesity and diabetes (O'Connor and Chapin 2003; Vom-Saal and Myers 2008).

BPA is not a monomer for PET. However, recent studies have reported BPA leaching from PET water bottles (Casajuana and Lacorte 2003; Santhi et al. 2012; Toyo'oka and Oshige 2000). It has been suggested that PET water bottles can be contaminated with BPA during recycling (Sax 2010). Another source of BPA in PET-bottled water could be due to bottle closures (Guart et al. 2011) or water itself (polluted prior to bottling) (Li et al. 2010). As a precautionary measure, in 2008, in the U.S. the use of BPA in children's products has been banned at the state and federal level (Erler and Novak 2010). On the other hand, the EU Regulation No 10/2011 has set a specific migration limit (SML) for BPA to food at 600 ng/g, and a total daily intake (TDI) for BPA of 0.05 mg/kg/body weight has been established (European Commission 2011).

References

Alin J, Hakkarainen M (2011) Microwave heating causes rapid degradation of antioxidants in polypropylene packaging, leading to greatly increased specific migration to food simulants as shown by ESI-MS and GC-MS. J Agric Food Chem 59(10):5418–5427. https://doi.org/10.1021/jf1048639

Al-Mudhaf HF, Alsharifi FA, Abu-Shady AS (2009) A survey of organic contaminants in household and bottled drinking waters in Kuwait. Sci Total Environ 407(5):1658–1668. https://doi.org/10.1016/j.scitotenv.2008.10.057

Anonymous (2013) Abietic acid. In: Hartwig A (ed) The MAK-Collection for occupational health and safety. Wiley-VCH Verlag GmbH & Co. KGaA, Weinheim. https://doi.org/10.1002/3527600418.mb51410kske3413

Arvanitoyannis IS, Kotsanopoulos KV (2014) Migration phenomenon in food packaging, food–package interactions, mechanisms, types of migrants, testing and relative legislation—a review. Food Bioprocess Technol 7(1):21–36. https://doi.org/10.1007/s11947-013-1106-8

ATSDR (1990) Toxicological profile for Di-n-butyl phthalate. Agency for Toxic Substances and Disease Registry (ATSDR), U.S. Department of Health and Human Services, Public Health Service, Atlanta

ATSDR (1993a) Toxicological profile for Di(2-ethylhexyl) phthalate. Agency for Toxic Substances and Disease Registry, U.S. Department of Health and Human Services, Public Health Service, Atlanta

ATSDR (1993b) Toxicological profile for diethyl phthalate. Agency for Toxic Substances and Disease Registry, U.S. Department of Health and Human Services, Public Health Service, Atlanta

Aznar M, Vera P, Canellas E, Nerín C, Mercea P, Störmerc A (2011) Composition of the adhesives used in food packaging multilayer materials and migration studies from packaging to food. J Mater Chem 21:4358–3470. https://doi.org/10.1039/C0JM04136J

Bach C, Dauchy W, Severin I, Munoz J-F, Etienne S, Chagnon MC (2013) Effect of temperature on the release of intentionally and non-intentionally added substances from polyethylene terephthalate (PET) bottles into water: chemical analysis and potential toxicity. Food Chem 139(1–4):672–680. https://doi.org/10.1016/j.foodchem.2013.01.046

Bach C, Dauchy W, Severin I, Munoz J-F, Etienne S, Chagnon MC (2014) Effect of sunlight exposure on the release of intentionally and/or non-intentionally added substances from polyethylene terephthalate (PET) bottles into water: chemical analysis and in vitro toxicity. Food Chem 162:63–71. https://doi.org/10.1016/j.foodchem.2014.04.020

Bach C, Dauchy X, David L, Etienne S (2012) Chemical compounds and toxicological assessments of drinking water stored in polyethylene terephthalate (PET) bottles: a source of controversy reviewed. Water Res 46(3):571–583. https://doi.org/10.1016/j.watres.2011.11.062

Becher G, Ovrum NM, Christman RF (1992) Novel chlorination by-products of aquatic humic substances. Sci Total Environ 117–118:509–520

Bentayeb K, Battle R, Romero J, Nerín C (2007) UPLC-MS as a powerful technique for screening the non-volitile contaminants in recycled PET. Anal Bioanal Chem 388(5–6):1031–1038. https://doi.org/10.1007/s00216-007-1341-9

Brüschweiler BJ, Küng S, Bürgi D, Muralt L, Nyfeler E (2014) Identification of non-regulated aromatic amines of toxicological concern which can be cleaved from azodyes used in clothing textiles. Regul Toxicol Pharmacol 69(2):263–272. https://doi.org/10.1016/j.yrtph.2014.04.011

Canellas E, Nerìn C, Moore R, Silcock PJ (2010) New UPLC coupled to mass spectrometry approaches for screening of non-volatile compounds as potential migrants from adhesives used in food packaging materials. Anal Chim Acta 666(1–2):62–69. https://doi.org/10.1016/j.aca.2010.03.032

Casajuana N, Lacorte S (2003) Presence and release of phthalic esters and other endocrine disrupting compounds in drinking water. Chromatographia 57(9–10):649–655. https://doi.org/10.1007/BF02491744

Cheng X, Shi H, Adams CD, Ma Y (2010) Assessment of metal contaminations leaching out from recycling plastic bottles upon treatments. Environ Sci Pollut Res 17(7):1323–1330. https://doi.org/10.1007/s11356-010-0312-4

Clayton GD, Clayton FE (1981) Patty's industrial hygiene and toxicology, vol. II A, 3rd edn. Wiley, Hoboken

Dabrowska A, Borcz A, Nawrocki J (2004) Aldehyde contamination of mineral water stored in PET bottles. Food Addit Contam 20(12):1170–1177. https://doi.org/10.1080/02652030310001620441

Di Felice R, Cazzola D, Cobror S, Oriani L (2008) Oxygen Permeation in PET Bottles with Passive and Active Walls. Packag Technol Sci 21:405–415

Dzieciol M, Trzeszczynski J (1998) Studies of temperature influence on volatile thermal degradation products of poly(ethylene terephthalate). J Appl Polymer Sci 69(12):2377–2381. https://doi.org/10.1002/(SICI)1097-4628(19980919)69:12%3C2377:AID-APP9%3E3.0.CO;2-5

EFSA Panel on Food Contact Materials, Enzymes and Processing Aids (CEP), Silano V, Barat Baviera JM, Bolognesi C, Brüschweiler BJ,Chesson A, Cocconcelli PS, Crebelli R, Gott DM, Grob K, Lampi E, Mortensen A, Rivière G, Steffensen IL, Tlustos C, Van Loveren H, Vernis L, Zorn H, Cravedi JP, Fortes C, de Fatima Tavares Poças M, Waalkens-Berendsen I, Wölfle D, Arcella D, Cascio C, Castoldi AF, Volk K, Cara-Carmona J, Castle L (2019) Draft update of the risk assessment of di-butylphthalate 2 (DBP), butyl-benzyl-phthalate (BBP), bis(2- 3 ethyl-hexyl)phthalate (DEHP), di-isononylphthalate (DINP) 4 and di-isodecylphthalate (DIDP) for use in food contact 5 materials. EFSA J. (In press). https://doi.org/10.2903/j.efsa.20YY.NNNN

EFSA Panel on Food Contact Materials, Enzymes, Flavourings and Processing Aids (2012) Scientific opinion on the safety evaluation of substance, titanium, nitride, nanoparticles, for as in food contact materials. EFSA J 10:2641–2649. https://doi.org/10.2903/j.efsa.2012.2641

Elobeid M, Allison D (2008) Putative endocrine disruptors and obesity: a review. Curr Opin Endocrinol Diabetes Obes 15(5):403–408. https://doi.org/10.1097/MED.0b013e32830ce95c

Erler C, Novak J (2010) Bisphenol A exposure: human risk and health policy. J Pediatr Nurs 25(5):400–407. https://doi.org/10.1016/j.pedn.2009.05.006

European Commission (2001) Regulation No. 262/2001/CE. Communication on the implementation of the community strategy for endocrine disruptors—arrange of substances suspected of interfering with the hormone systems of humans and wildlife. Off J Eur Comm L39:16–17

European Commission (2007) Commission Directive 2007/19/EC of 2 April 2007 amending Directive 2002/72/EC relating to plastic materials and articles intended to come into contact with food. European Union, Bruxelles. Off J Eur Union L91:17–36

European Commission (2011) Commission Regulation (EU) No 10/2011 of 14 January 2011 on plastic materials and articles intended to come into contact with food. Off J Eur Union L12:1–89

European Parliament and Council (2004) Regulation (EC) No 1935/2004 of the European Parliament and of the Council of 27 October 2004 on materials and articles intended to come into contact with food and repealing Directives 80/590/EEC and 89/109/EEC. Off J Eur Union L338:4–17

Freire A, Castle L, Reyes FGR, Damant AP (1998) Thermal stability of polyethylene terephthalate food contact materials: formulation of volatiles from retail samples and implications for recycling. Food Addit Contam 15(4):473–480

Ghanem A, Maaloly J, Saad RA, Salameh D (2013) Saliba CO (2013) Safety of Lebanese bottled waters: VOCs analysis and migration studies. Am J Anal Chem 4:176–189. https://doi.org/10.4236/ajac.2013.44023

Glaze WH, Koga M, Cancilla D (1989) Ozonation by-products. Improvement of an aqueous phase derivatization method for the detection of formaldehyde and other carbonyl compounds formed by the ozonation of drinking water. Environ Sci Technol 23(7):838–847. https://doi.org/10.1021/es00065a013

Godish T, Davis WT, Fu JS (2015) Air Quality, 5th edn. CRC Press, Boca Raton

Gomez-Hens A, Aguilar-Caballos MP (2003) Social and economic interest in the control of phthalic acid esters. Trends Anal Chem 22:847–857

Gramshaw JW, Vandenburg HJ, Lakin RA (1995) Identification of potential migrants from sample of dual-ovenable plastics. Food Addit Contam 12:211–222

Greifenstein M, White DW, Stubner A, Hout JJ, Whelton A (2013) Impact of temperature and storage duration on the chemical and odor quality of military packaged water in polyethylene terephthalate bottles. Sci Total Environ 456–457C:376–383. https://doi.org/10.1016/j.scitotenv.2013.03.092

Grob K, Biedermann M, Scherbaum G, Roth M, Rieger K (2006) Food contamination with organic materials in perspective: packaging materials as the largest and least controlled source? A view focusing on the European situation. Crit Rev Food Sci Nutr 46(7):529–536. https://doi.org/10.1080/10408390500295490

Guart A, Bono-Blay F, Borrell A, Lacorte S (2011) Migration of plasticizers phthalates, bisphenol A and alkylphenols from plastic containers and evaluation of risk. Food Addit Contam Part A Chem Anal Control Expo Risk Assess 28(5):676–685. https://doi.org/10.1080/19440049.2011.555845

Harvey JA (2012) Chemical and physical aging of plastics. In: Kutz M (ed) Handbook of environmental degradation of materials. Delmar, New York

Hollnagel HM, van Herwijnen P, Sura P (2014) Assessing safety of non-intentionally added substances in polymers used for food contact applications. Toxicol Lett 229:S22–S39. https://doi.org/10.1016/j.toxlet.2014.06.158

IARC (1982) Monographs on the evaluation of carcinogenic risks to humans, 29, 257: Suppl. 7, 62. International Agency for Research on Cancer (IARC), Lyon

IARC (2006) Monographs on the evaluation of carcinogenic risks to humans. Formaldehyde, 2-Butoxyethanol and 1-tert-Butoxypropan-2-ol, vol 88. International Agency for Research on Cancer (IARC), Lyon

Ikem A (2010) Measurement of volatile organic compounds in bottled and tap waters by purge and trap GC-MS: are drinking water types different? J Food Comp Anal 23(1):70–77. https://doi.org/10.1016/j.jfca.2009.05.005

ILSI (2000) Packaging materials: 1. Polyethylene Terephthalate (PET) for food packaging applications. International Life Science Institute (ILSI), Brussels

Ji J, Deng C, Shen W, Zhang X (2006) Field analysis of Benzene, Toluene, Ethylbenzene and Xylene in water by portable gas chromatography-microflame ionization detector combined with headspace solid-phase microextraction. Talanta 69(4):894–899. https://doi.org/10.1016/j.talanta.2005.11.032

Keresztes S, Tatàr E, Mihucz VG, Viràg I, Majdik C, Zàray G (2009) Leaching of antimony from polyethylene terephthalate (PET) bottles into mineral water. Sci Total Environ 407(16):4731–4735. https://doi.org/10.1016/j.scitotenv.2009.04.025

Kim H, Gilbert SG, Johnson JB (1990) Determination of potential migrants from commercial amber polyethylene terephthalate bottle wall. Pharm Res 7(2):176–179. https://doi.org/10.1023/A:1015884920237

Krochta JM (2007) Food packaging. In: Heldman DR, Lund DB (eds) Handbook of food engineering, 2nd edn. Taylor and Francis, Abingdon-on-Thames

Li D, Han B, Liu Z, Zhao D (2001) Phase behavior of supercritical CO2/styrene/poly(ethylene terephthlate) (PET) system and preparation of polystyrene/PET composites. Polymer 42(6):2331–2337. https://doi.org/10.1016/S0032-3861(00)00601-7

Li X, Ying GG, Su HC, Yang XB, Wang L (2010) Simultaneous determination and assessment of 4-nonylphenol, bisphenol A and triclosan in tap water, bottled water and baby bottles. Environ Int 36(6):557–562. https://doi.org/10.1016/j.envint.2010.04.009

Liteplo RG, Beauchamp R, Meek ME, Chénier R (2002) Formaldehyde. Concise International Chemical Assessment Document No. 40. World Health Organization, Geneva. https://www.who.int/ipcs/publications/cicad/en/cicad40.pdf. Accessed 21 Nov 2019

Loos R, Hanke G, Umlauf G, Eisenreich SJ (2007) LC-MS-MS analysis and occurrence of octyl- and nonylphenol, their ethoxylates and their carboxylates in Belgian and Italian textile industry, waste water treatment plant effluents and surface waters. Chemosphere 66(4):690–699. https://doi.org/10.1016/j.chemosphere.2006.07.060

Ma TH, Harris MM (1988) Review of the genotoxicity of formaldehyde. Mutat Res 196:37–59

McNeal TP, Biles JE, Begley TH, Craun JC, Hopper ML, Sack CA (2000) Determination of suspected endocrine disruptors in foods and food packaging. ACS Symp Ser 747:33–52

Migliore L, Ventura L, Barale R, Loprieno N, Castellino S, Pulci R (1989) Micronuclei and nuclear anomalies induced in the gastro-intestinal epithelium of rats treated with formaldehyde. Mutagenesis 4(5):327–334. https://doi.org/10.1093/mutage/4.5.327

Morelli-Cardoso MHW, Tabak D, Cardoso JN, Pereira AS (1997) Application of capillary gas chromatography to the determination of ethylene glycol migration from PET bottles in Brazil. J High Resolut Chrom 20(3):183–185. https://doi.org/10.1002/jhrc.1240200313

Muncke J (2009) Exposure to endocrine disrupting compounds via the food chain: is packaging a relevant source? Sci Total Environ 407(16):4549–4559. https://doi.org/10.1016/j.scitotenv.2009.05.006

Mutsuga M, Kawamura Y, Sugita-Konishi Y, Hara-Kudo Y, Takatori K, Tanamoto K (2006) Migration of formaldehyde and acetaldehyde into mineral water in polyethylene terephthalate (PET) bottles. Food Add Contam 23(2):212–218. https://doi.org/10.1080/02652030500398361

Nawrocki J, Dabrowska A, Borcz A (2002) Investigation of carbonyl compounds in bottled waters from Poland. Water Res 36(19):4893–4901. https://doi.org/10.1016/S0043-1354(02)00201-4

Nerin C, Alfaro P, Aznar M, Domeño C (2013) The challenge of identifying non-intentionally added substances from food packaging materials: a review. Anal Chim Acta 775:14–24. https://doi.org/10.1016/j.aca.2013.02.028

O'Connor JC, Chapin RE (2003) Critical evaluation of observed adverse effects of endocrine active substances on reproduction and development, the immune system, and the nervous system. Pure Appl Chem 75(11–12):2099–2123. https://doi.org/10.1351/pac200375112099

Osorio J, Ubeda S, Aznar M, Nerìn C (2018) Analysis of isophthalaldehyde in migration samples from polyethylene terephthalate packaging. Food Addit Contam Part A Chem Anal Control Expo Risk Assess 35(8):1645–1652. https://doi.org/10.1080/19440049.2018.1465208

Overman DO (1985) Absence of embryotoxic effects of formaldehyde after percutaneous exposure in hamsters. Toxicol Lett 24(1):107–110

Ozlem KE (2008) Acetaldehyde migration from polyethylene terephthalate bottles into carbonated beverages in Turkiye. Int J Food Sci Technol 43(2):333–338. https://doi.org/10.1111/j.1365-2621.2006.01443.x

Pezo D, Fedeli M, Bosetti O, Nerìn C (2012) Aromatic amines from polyurethane adhesives in food packaging: the challenge of identification and pattern recognition using Quadrupole-Time of Flight-Mass Spectrometry[E]. Anal Chim Acta 756:49–59. https://doi.org/10.1016/j.aca.2012.10.031

Romão W, Franco MF, Corilo YE, Eberlin MN, Spinacé MAS, De Paoli MA (2009) Poly(ethylene terephthalate) thermo-mechanical and thermo-oxidative degradation mechanisms. Polymer Degrad Stabil 94(10):1849–1859. https://doi.org/10.1016/j.polymdegradstab.2009.05.017

Salaspuro M (2011) Acetaldehyde and gastric cancer. J Digest Dis 12:51–59

Santhi VA, Sakai N, Ahmad ED, Mustafa AM (2012) Occurrence of bisphenol A in surface water, drinking water and plasma from Malaysia with exposure assessment from consumption of drinking water. Sci Total Environ 427–428:332–338. https://doi.org/10.1016/j.scitotenv.2012.04.041

Sax L (2010) Polyethylene terephthalate may yield endocrine disruptors. Environ Health Perspect 118(4):445–448. https://doi.org/10.1289/ehp.0901253

Selke S (2005) Food packaging: plastics. In: Hui Y (ed) Handbook of food technology and engineering, vol 3. CRC, Boca Raton

Serrano SE, Braun J, Trasande L, Dills R, Sathyanarayana S (2014) Phthalates and diet: a review of the food monitoring and epidemiology data. Environ Health 13(1):43. https://doi.org/10.1186/1476-069X-13-43

Sheftel VO (2000) Indirect food additives and polymers: migration and toxicology. American Chemical Society, Boca Raton

Swan SH, Main KM, Liu F, Stewart SL, Kruse RL, Calafat AM, Mao CS, Redmon JB, Ternand CL, Sullivan S, Teague JL (2005) Study for future families research team. Decrease in anogenital distance among male infants with prenatal phthalate exposure. Environ Health Perspect 113(8):1056–1061. https://doi.org/10.1289/ehp.8100

Toyo'oka T, Oshige Y (2000) Determination of alkylphenols in mineral water contained in PET bottles by liquid chromatography with coulometric detection. Anal Sci 16(10):1071–1076. doi:10.2116/analsci.16.1071

US-EPA (2009) National primary drinking water regulations. http://www.epa.gov/ogwdw/consumer/pdf/mcl.pdf. Accessed 21 Nov 2019

Vom-Saal FS, Myers JP (2008) Bisphenol A and risk of metabolic disorders. JAMA 300(11):1353–1355. https://doi.org/10.1001/jama.300.11.1353

Welle F (2011) Twenty years of PET bottle to bottle recycling—an overview. Resour Conserv Recycl 55(11):865–875. https://doi.org/10.1016/j.resconrec.2011.04.009

Welle F, Franz R (2011) Migration of antimony from PET bottles to beverages: determination of the activation energy of diffusion and migration modeling compared with literature data. Food Addit Contam Part A Chem Anal Control Expo Risk Assess 28(1):115–126. https://doi.org/10.1080/19440049.2010.530296

Zhang H, Ward IM (1995) Kinetics of hydrolytic degradation of poly(ethylene naphtalene-2,6-dicarboxylate). Macromolecules 28(23):7622–7629. https://doi.org/10.1021/ma00127a006

Zoeller RT, Brown TR, Doan LL, Gore AC, Skakkebaek NE, Soto AM, Woodruff TJ, Vom Saal FS (2012) Endocrine-disrupting chemicals and public health protection: a statement of principles from The Endocrine Society. Endocrinol 153(9):4097–4110. https://doi.org/10.1210/en.2012-1422

Printed in the United States
By Bookmasters